Daniel Pacheco
Victor Noble

Production scheduling for parallel machines:

Daniel Pacheco
Victor Noble

Production scheduling for parallel machines:

Hybrid methods for production environments with sequence-dependent enlistment times

ScienciaScripts

Imprint

Cover image: www.ingimage.com

This book is a translation from the original published under ISBN 978-613-9-00752-3.

Publisher:
Sciencia Scripts
is a trademark of
Dodo Books Indian Ocean Ltd. and OmniScriptum S.R.L publishing group

120 High Road, East Finchley, London, N2 9ED, United Kingdom
Str. Armeneasca 28/1, office 1, Chisinau MD-2012, Republic of Moldova, Europe
Managing Directors: Ieva Konstantinova, Victoria Ursu
info@omniscriptum.com

Printed at: see last page
ISBN: 978-620-8-37671-0

Contents

DEDICATION

With deep gratitude, I dedicate this research work to God, the supreme source of our inspiration and strength. His guidance has been the light that led us successfully to the culmination of this academic project, representing one of our greatest desires.

With all my heart, I wish to dedicate this achievement in a special way to my dearest best friend, **Armando Peña.** As a fellow university student and confidant, his memory lives on in our hearts with love and gratitude. Despite his departure to heaven, a victim of relentless cancer,

his courageous spirit, charisma and unwavering friendship leave an indelible mark on this work. May this joint effort serve as a heartfelt tribute to his memory and to the enduring friendship that dwells eternally in our hearts.

CHAPTER 1

INTRODUCTION

In most companies their competitiveness is based on their prices, one of the options to achieve this objective is to make use of production planning and control (PCP). This area helps to plan, determine, make decisions and identify processes, in order to determine, for example, which are the aspects in which the company incurs more expenses, in such a way that an analysis and/or improvement can be made to minimise operational costs, so that the company can have greater productivity without neglecting the quality of the products (Prado Bustamante, 1992).

The most studied PCP problems are *lot sizing* and production *sequencing,* where the purpose of the first case is to provide a production strategy that can determine in what time and in what quantity each product should be produced, taking into account the changing demands over time (Padrón Cano, 2012).

Likewise, the sequencing of production is done in order to facilitate and avoid direct and indirect costs that may be incurred in any scenario, as could be the case of the costs of preparing each product, costs of machine openings, among others. These decisions are made based on the needs of each organisation. Taking as an example an industry that requires a production sequence and has only one machine, it will have to know that for each period a sequencing of the products handled by the company must be calculated (Valero, Sabater, Bas, and Díaz, 2001).

Decisions on these problems are traditionally made separately, but it has been shown that in this way decisions can be suboptimal or become unfeasible (Gómez Urrutia, Aggoune, and Dauzère-Pérès, 2014), which is why integrated problems are currently being studied, so that most companies, whether small, medium or large, are able to identify and adequately apply methods that help them to effectively solve their production scheduling.

When the integrated production sizing and sequencing problem is answered, benefits are obtained over the entire planning horizon, as a good production plan must be able to intelligently sequence product batches and satisfy a variety of conditions in order to be realistic. That is why this problem can also be seen in two-level synchronised production environments, which can be observed for example in beverage production, foundries, grain, glass container industry, animal feed production, pharmaceutical company, sand casting operations, among others (Ferreira, Clark, Almada-Lobo, and Morabito, 2012).

The planning horizon mentioned above is often time-based, making it a very important factor for companies to make timely decisions at certain critical and/or analytical moments of an organisation. This planning horizon can be formulated within the scope of long-, medium- and short-term plans, which are known as the decision types of an enterprise, as explained in Figure 1.1.

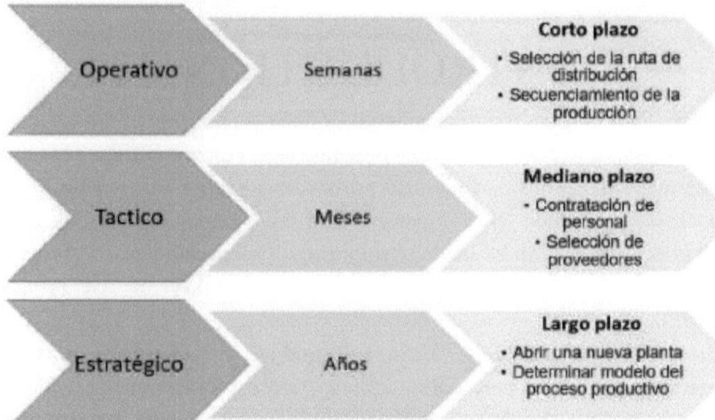

Figure 1.1: Decision types in production environments.
Source: Own elaboration

There are particular production decisions at the operational level such as determining the order of production, number of products and quantities to be produced on each machine. At the beginning of the study of these decisions, they used to be determined separately (Boonmee and Sethanan, 2016). For example, on the one hand, the order of production was determined, on the other hand, the quantities to be produced and the allocation to the different machines that make up the factory floor.

However, these decisions present a *trade off* between sequencing and sizing, on the one hand if emphasis is placed on sequencing, fewer but large batches are produced, generating high inventory levels, on the other hand if sizing is prioritised, small batches are produced, improving the inventory cost but increasing the number of batches, thus increasing the number of exchanges between products and therefore the *setup* cost (Noble Ramos, 2017).

The research topic developed in this work is a complex problem that encompasses precisely the two aspects mentioned, batch sizing and production sequencing, this problem is considered complex because it is a combinatorial problem with exponential growth as its size increases.

To solve this problem, 2 types of approaches can be used: exact or approximate, within the exact approach the most used algorithms are *Branch and Bound,* cutting planes, among others (Pousa, 2013). While heuristics and metaheuristics are part of the approximate approach.

Considering the above, in the present research a set of approximate methods (metaheuristics) was developed to provide an answer to the integrated problem of batch sizing and production sequencing in parallel multi-machine production environments with sequence-dependent *setups* (PMSD-LSP), although it is true that exact methods are remarkable because they find an optimal solution, generally these methods tend to consume a lot of computational time, so metaheuristics become an alternative for such a case, because they are not concerned with obtaining the best solution, but to obtain a good solution in a reasonable time, thus allowing to achieve the proposed organisational objectives (Marti, 2003).

Now, in order to obtain good results both in terms of solution quality and computational time, a hybrid approach was developed for the proposed metaheuristics, so that the allocation and sequencing of production arises from the adaptation of metaheuristics, while an exact model is responsible for batch sizing and evaluation of the solution, it should be noted that these two parts work together, i.e. both depend on each other to function properly. This method aims to obtain a feasible solution to this problem, where it manages to determine the sequencing of production, the amount of products to be made in certain periods for each machine and each product, in addition to the decision of how much will be inventoried or delayed depending on the conditions that may favour the production environment, in order to meet the requirements and changing needs with respect to time.

CHAPTER 2

PROBLEM STATEMENT

2.1. Description of the problem

The integrated problem of batch sizing and sequencing of production with parallel machines and sequence-dependent *setups* (PMSD-LSP) is responsible for determining the quantities to be produced, inventoried and the quantities to be backlogged, taking into account the capacity of each company. At the same time, the allocation of these products to different machines and the order in which they will be produced is also studied. All this is done with the aim of reducing the total cost associated with these decisions.

Multiple organisations run their production layout with a configuration of multiple machines working in parallel, shown in Figure 2.1. Products may be assigned to some or all of these machines depending on the production plan. There may be *setup* times, which correspond to the time spent on adjustments, preparation of the machines and/or raw materials among others, and which are performed before starting the production operation. This type of production system is the one studied in this work.

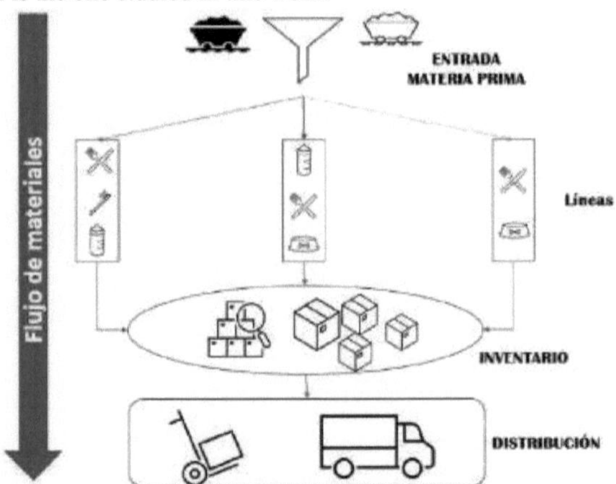

Figure 2.1: Configuration of a production system of multiple machines working in parallel.
Source: Own elaboration

From Figure 2.1 presented above, it can be seen that there are several machines producing different products. It is assumed that all machines are capable of making all products. However, in some production plan a product may be destined to be produced on only one of the machines (as long as only one batch is made in that period) or on several machines. Finally it goes to a storage stage and after that to the respective distribution. The allocation can best be analysed in Table 2.1.

	Sequence			
Machines				
	Articles	MY	M2	M3
	Al	1	3	
	A2	2		

A3	3	2	2
A4		1	1

Table 2.1: Sequencing of products in machines

Source: Own elaboration

Table 2.1 shows that there are 4 items and 3 machines. Each machine can make different products, i.e. a product can be made on any of the machines. In the example, note that product A4 was not assigned to machine MI, which means that this product cannot be produced on that machine, because the optimal production schedule indicates this. However, assigning a product to all machines does not guarantee its production on all machines, due to the capacity criteria that must be met.

Also from Table 2.1 we can say that there is an order (sequencing) of the products for each machine. For example, for machine M2, first the product A4 will be produced, secondly the product A3 and lastly the product Al.

When changing products on the machines, there are sequence-dependent *setups*, which leads to longer production times. This means that, for example, it is not the same to produce a sugary soft drink first and then a *light* soft drink, as it is to produce them in the reverse order, since going from a sugary soft drink to a light soft drink takes longer than going from a *light soft drink* to a sugary soft drink, it should be noted that there may be products that go through some machines or products that have to be made after a certain product.

For example, each machine has a sequence to meet the target, the MI in its production plan indicates that it must first produce the Al, then the A2 and finally, the A3. There are also exceptions mentioned above where some items are not produced on some machines, such as A4, which cannot be produced on the IM.

Production decisions related to sizing and sequencing must be made taking into account all these aspects. Also the estimation of the costs related to sizing (inventory and backlog costs) and those related to sequencing (*setup* costs) must be done conscientiously. Since a wrong determination of these costs will unbalance the decision making in favour of one or the other objective.

To carry out planning at the operational level, various techniques are used, such as heuristic planning based on experience, simulation, among others. In some cases, mathematical models are used or different algorithms are programmed that help to obtain optimal and very beneficial results when planning production. These models and algorithms have been modified and adapted to different problems that arise over time.

Of the mathematical models that have emerged to solve the PMSD-LSP, the following stand out: *The general lot sizing and scheduling problem* (GLSP), *The discrete lot sizing and scheduling problem* (DLSP), *The continuous setup lot sizing problem* (CSLP), *Capacited LotSizing Problem with Sequence Dependent Setup Cost* (CLSD), among others. All these models are born from a basic model which is *The capacitated lot sizing problem* (CLSP), being this a model intended only for sizing, from this the other models arise as formulated extensions, where new features are included to the base problem to integrate sequencing (Noble Ramos, 2017). The CLSP considers planning periods in which multiple products can be manufactured, the problem seeks to determine how much to produce in each period in such a way that orders with delivery times prior to the end of that period are fulfilled in such a way as to minimise the costs incurred (Absi, 2008). Multiple researchers and practitioners have made improvements to adapt it to different situations, including variable machine capacity, consideration of workers, among others (Karimi, Ghomi, and Wilson, 2003).

Although these models are very efficient, since they are able to obtain the global optimum, they are not so efficient when it comes to computational time, which can generate a considerable computational expense for real size problems, this is due to the fact that these types of models have a degree of *NP-HARD* complexity which implies the creation of millions of variables giving way to the generation of thousands of constraints (Bonrostro, 2015).

Due to the computational inefficiency that can occur in the resolution by exact methods, the programming of different algorithms (Heuristics, Metaheuristics, among others) that are more efficient in computational time and that generate a feasible response or even an optimal result according to the flexibility and robustness of the algorithm is justified.

Given the existence of mathematical models and approximate solution methods where adaptive techniques such as metaheuristics and heuristics can be found to solve the production sizing and sequencing problem, it is natural to ask about the efficiency of these techniques under particular solution conditions.

This is because most of the research consulted so far only focuses on one of the two approaches: the exact or approximate resolution, where the approximate ones can be the heuristics that are based on the study of exact or precise solutions, while the metaheuristics that are based on the search for not only exact solutions are also based on the search for approximate solutions (Maldonado, 2016).

2.2. Problem formulation

From what has been described above, the following problem question is proposed: How to solve by means of approximate solution methods the integrated problem of batch sizing and production sequencing in parallel multi-machine production environments with sequence-dependent *setups*?

CHAPTER 3

OBJECTIVES

3.1. General Objective

To compare, through a descriptive analysis, the performance of different hybrid algorithms for the integrated problem of batch sizing and production sequencing in parallel multi-machine production environments with sequence-dependent *setups*.

3.2. Specific Objectives

I. Analyse existing and proposed algorithms that can be implemented to solve the integrated problem of batch sizing and production sequencing in parallel multi-machine production environments with sequence-dependent *setups*.

II. Adapt different approximate solution methods chosen for the integrated problem of batch sizing and production sequencing in parallel multi-machine production environments with sequence-dependent *setups*.

III. Parameterise the chosen algorithms, if necessary, for their subsequent application.

IV. Contrast by descriptive analysis the quality of the solution and the response time for each algorithm presented.

CHAPTER 4

FRAME OF REFERENCE

Conceptual Framework

Operations Research

Operations research usually involves the use of mathematical models, statistics and algorithms to model and solve complex problems, determining the optimal solution and thus enabling decisions to be made. It often involves the study of complex real systems, with the aim of improving (or optimising) their performance. Operations research allows the analysis of decision making taking into account the scarcity of resources, to determine how a defined objective, such as profit maximisation or cost minimisation, can be optimised (Castillo, Conejo, Pedregal, García, and Alguacil, 2002).

Mathematical Programming

Mathematical programming "is a powerful optimisation technique used in the decision-making process of many organisations, it uses models to represent those aspects of reality that have an influence on its field of interest, in this case the decisions that optimise the operation of a system" (UNED, 2017).

Within the problems that concern mathematical programming, there is always an objective function that we seek to maximise or minimise, this will depend on the context in which we are working, this function will be subject to a set of conditions, or in this case restrictions, which must necessarily be met (Castillo and cois., 2002).

Algorithm

In mathematics, logic, computer science and related disciplines, an algorithm is a set of well-defined, ordered and constrained instructions or rules that generally allow you to solve problems, perform calculations, process data and perform other tasks or activities. Given the initial state and input, follow successive steps to reach the final state and obtain a solution. Algorithm consists of the search for solutions to a specific problem (Vancells Flotats, 2002).

Algorithms can be found in both basic and complex activities. An algorithm is a cooking recipe, a division algorithm, a Monte Carlo method or a fast Fourier transform. We understand an algorithm as a structure of actions set up to perform tasks. Following an algorithm is not the same as building an algorithm, because the latter must understand how to perform the tasks in depth to decompose them into actions and reintegrate them to build them (Sepúlveda, Díaz, and Arrieta, 2014).

Metaheuristics

Metaheuristic procedures or algorithms are a class of approximate methods that are designed to solve difficult combinatorial optimisation problems where classical heuristics are neither effective nor efficient. Metaheuristics provide a general framework for creating new hybrid algorithms by combining different concepts derived from artificial intelligence, biological evolution and statistical mechanics (Vélez and Montoya, 2007).

Heuristics

According to Pacheco Bonrostro (n.d.), heuristics are "solution methods that do not ensure obtaining the optimal solution, but a solution that is usually close to the optimal one, in a shorter computation time (which normally grows polynomially depending on the number of variables)". Through this concept, people are able to test their knowledge and creativity, designing methods and strategies that help to solve certain problems quickly, problems that, if studied properly, giving compliance to conditions and needs that are present at any given time, can yield very reliable solutions that favour organisations that present urgent uncertainties throughout their functional process (Bonrostro, Bonrostro, Bonrostro, Bonrostro, Bonrostro, Bonrostro, Bonrostro and Bonrostro, 2001).Bonrostro, 2015).

4.2. Theoretical framework

Planning y Production Control

Within this concept is based largely on what can be the profits of many organisations, proper production planning can mean saving many resources, and if this is not properly controlled, errors will begin to stand out that would cost the confidence of the organisation (Villay Pereira et al., 2013)..

Production planning and control (PCP) begins by establishing the objective of the production system, defining the resources, personnel and equipment that will be needed to meet the demands that arise over a specific period, these demands may be changing throughout the period, and it is through a good PCP defined by the company, that it can meet deliveries on time, minimizing what would be costs of delays or unnecessary inventories (Mesquita, 2017).Mesquita, 2017).

There are two fundamental aspects to which the PCP must give priority. The first is the sizing of the batches that will be in the production environment, i.e. how many units to produce in each batch, depending on the maximum time available for the product to be worked and thus the maximum capacity that can be produced, The number of machines used also influences the right batch sizing decision, as does the number of products being produced, and depending

on whether the expected demand in the next period is expected to be higher than the current one, the configuration of the batch sizing may be different, in order to meet the demand and not incur additional costs (Azocar and Hornig, 2014).
The second aspect necessary for good production planning and control deals with the sequencing of production, i.e. in what order the batches should be produced, within this it is important to take into account the processing times of each batch and the lead time when a production change is incurred, the ideal is to find the best sequence of processing with the intention of facilitating production within the system and expanding capacity within each period.

Solution methods

The integrated problem of batch sizing and production sequencing is a peculiar problem, in this type of investigations one seeks to optimise one or more objectives. Among these problems, the variables grow significantly as the scale of the problem increases. Therefore, the decision maker must have tools to help him in his decision. For this type of problem, different solutions are applied to try to solve the problem. In this paper, these methods are divided into two types: exact methods and approximate methods.

I. **Exact Methods**

The exact method is the method of trying to obtain the best overall solution regardless of the time and resources invested in solving the problem. Some particular exact methods for a problem can be very efficient for relatively large problems, such as the Hungarian method for the assignment problem. Other exact methods can be flexible for various and varied problems but can only be used efficiently for problems whose size allows the best solution to be found in a reasonable time. Such is the case for solution techniques based on the simplex method. Other examples of exact methods include: linear programming, binary linear programming, Thangavelu method, dynamic programming (Bozer and Wang, 2012).

II. **Approximate Methods**

In recent years, approximate methods to study the problem of batch sizing and sequencing of production have become very important because they can be used as estimates for continuous and efficient improvement of the company (Matai, Singh, and Mittal, 2010).
Even if it cannot be proven to be optimal, approximate methods prove to be efficient procedures for finding a good solution. In these methods, the speed of processing is as important as the quality of the solution obtained (Puris, Hern, and Bayas, 2020). However, these two aspects will depend on a correct estimation of the method parameters and an efficient programming of the method.
Among these approximation methods, we can find mainly two types: heuristic and metaheuristic.

a) **Heuristics**

Heuristic methods are the process of solving well-defined optimisation problems using intuitive methods, where the structure of the problem is used intelligently to obtain a good solution. Heuristic algorithms can find "good" solutions with reasonable time and computing resources in larger problem cases, but cannot offer a guarantee for finding the best solution to the problem. They are based on the intuition of researchers and are usually developed for specific problems (Fernandez et al., 1996).
Heuristics is considered the art of human invention and its purpose is to find strategies, methods and standards that can solve problems through creativity, disagreement or lateral thinking. Similarly, it is true that the heuristic method relies on the individual's own

experience and the experience of others to find the most feasible solution (Huaracha Ortega, 2015).

For example, heuristics can be seen as a theory that stimulates the thinking of the people responsible for analysing all the materials collected during the research. In this sense, it is true that it is related to decision-making and can solve problems without ensuring that the options used are the most appropriate ones (Heuristics, n.d.).

Now, as a scientific discipline, heuristics can be applied to any science in the broadest sense to develop methods, principles, rules or strategies that help find the most effective and efficient solutions to individual analysis problems (Contreras, Sierra, Hernández, Hernández, & Moyotl, 2020).

Heuristic methods are a set of methods and techniques used to solve problems when it is difficult to find the best solution. Therefore, heuristic methods are often used in scientific disciplines to obtain the best results in a specific problem (Camelo Salinas, 2020).

These methods are not new, in fact heuristics have existed since ancient Greece, however, this term was popularised by the mathematician George Pólya, in his book "How to solve it", in which he explains the heuristic method to his mathematics students, and to all those who wish to learn the discipline, by means of four examples (Heuristics, n.d.):

- If a problem is not understood, a scheme is drawn.
- If you don't find the solution, you pretend you already have it and deduce from it (reasoning in reverse).
- If the problem is abstract, a concrete example is tested.
- A more general problem is addressed first and reviewed.

The philosopher and mathematician Lakato believes that heuristics are a set of methods or rules that can be affirmative or negative, and indicate which are the ideal measures to solve problems. Lakato pointed out in his work on scientific research projects that the structure of each project can be positively or negatively oriented (López-Jiménez, 2017).

With respect to the above, positive heuristics are to establish guidelines on how to develop research plans. In contrast, negative programme heuristics mean something that cannot be changed or modified. For example, in computer science, heuristics include finding or building algorithms with good execution speed, such as computer games or programs that detect whether emails are *spam* (López-Jiménez, 2017).

&) Metaheuristics

Metaheuristics are a heuristic method used to solve a class of general computational problems, using the parameters given by the users in some general and abstract processes by which it is expected to achieve effective results. In general, these approximate solution methods are given from the complexity of a given problem, knowing that the computational time needed to solve it through exact methods, is a very high time where too high expenses are incurred (Crainic, Davidovic, and Ramljak, 2014).

The best computational solution is difficult to obtain for many optimisation problems that are very important in industry and science. In fact, it is possible to be satisfied with the "good" solution provided by these approximate solution methods (Heuristics and Metaheuristics). Most metaheuristics are aimed at combinatorial optimisation problems, but they can of course be applied to any problem that can be reconstructed in heuristic terms (Crainic et al, 2014).

Combinatorial optimisation is very much in the area of applied mathematical optimisation which is related to operations research and computational complexity. Combinatorial optimisation algorithms solve problems that are generally difficult to solve by exploring the (usually large) solution space of these instances. Combinatorial optimisation algorithms achieve this goal by reducing the effective size of the space and efficiently exploring the search space (Marti, 2003).

Gogna and Tayal (2013) emphasise that approximate methods aim to provide high-quality solutions in a reasonable time (rather than guaranteeing a globally optimal solution). He further defines metaheuristics as an iterative generative process that guides subordinate heuristics by intelligently combining different concepts to explore and exploit the search space.

According to Abdel-Basset, Abdel-Fatah, and Sangaiah (2018), a metaheuristic is an algorithmic structure that is generally applied to several optimisation problems and can be adapted to a given problem with only minor modifications. Metaheuristics are general and can be applied to any problem of high complexity, guaranteeing their performance, unlike heuristics, which are only very effective when they are only applicable to a few types of problems.

However, Peralta-abarca and Moreno-bernal (2019) define metaheuristics as an advanced search process that applies one (or more) heuristic rules, which allows exploration of the search space rather than exhaustive search (in terms of time required and computational resources) and is more efficient (in terms of the quality of computational solutions) than a simple heuristic search. The effectiveness of metaheuristics depends on the ability to accommodate two key concepts when searching the solution space, these concepts are exploration (diversification) and exploitation (intensification). Figure 4.1 shows the behaviour in each case.

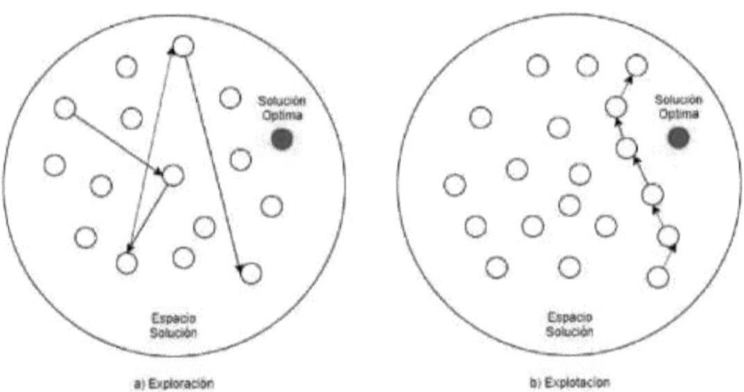

Figura 1: Exploration and intensification phases.
Source: Martinez Lopez (2018)

It can be said that the exploration process is the ability of mctahcuristics to traverse the solution space in order to avoid getting stuck in local optima. On the other hand, the exploitation process allows evaluating adjacent solutions from which the best solution is expected to be found (Martínez Lopez, 2018).

In general, a good mctahcuristic maintains the right balance of these concepts by exploring local escape optima when they are detected and reinforcing those concepts when the solution shows a high probability of approaching the optimal solution (Martínez Lopez, 2018).

However, metaheuristics are classified into categories and subcategories, showing the diversity of these solution methods. Figure 4.2 illustrates the taxonomy associated with metaheuristics.

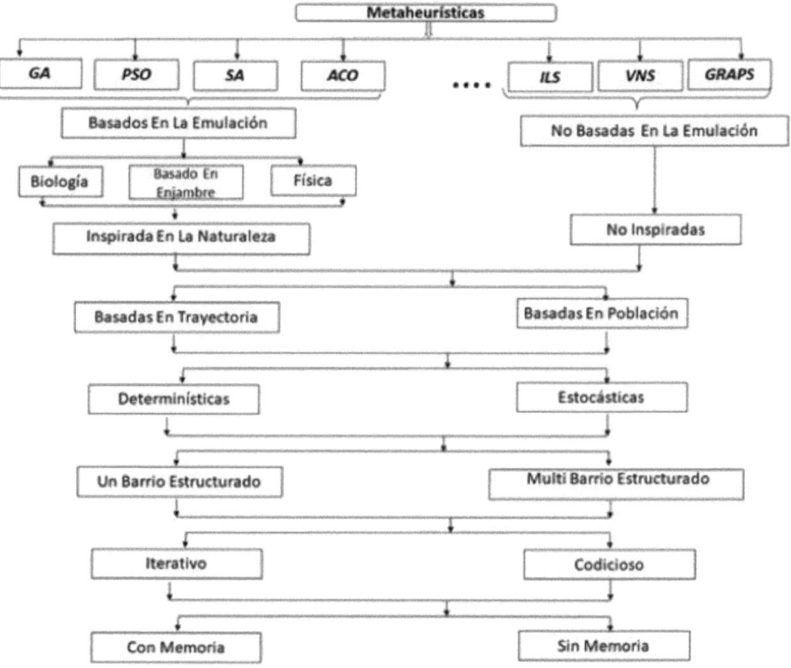

Figura 2: Taxonomy of Metaheuristics
Source: Abdel-Basset et al. (2018).

By analysing the presented figure we can observe that there are different types of meta-heuristics studied, among them, genetic algorithm (GA), particle swarm optimisation (PSO), simulated annealing (SA), ant colony optimisation (ACO), iterated local search (ILS), variable environment search (VXS) and greedy random adaptive search procedure (GRAPS). These meta-heuristics are classified as emulation-based or non-emulation-based. Emulation is the act of imitating something in order to match it. We can find in emulation-based Metaheuristics that are inspired by nature, from here it is decided whether they are trajectory-based or population-based, as well as analysing whether it is stochastic or deterministic. In this way the principles and classification methods for metaheuristics can be illustrated.

Genetic Algorithm

Genetic algorithms are evolutionary metaheuristics, often used in search and parameter optimisation problems, based on the principles of sexual reproduction and survival of the fittest (Fogel Pedroso, 2000) (Fogel, Hsu, Shapiro, Xelson-Goens, and Secrist, 2006). One of the first to contribute to this algorithm was Holland (1975), who developed a technique that mimicked natural selection in its execution.

These algorithms are divided into several stages, the most important of which are the selection, crossing and mutation stages. In the selection stage, the best individuals are chosen, which we will call parents, to later arrange for their crossing, as this is an algorithm that tries to imitate the behaviour of living beings, the parents selected are two, these will be the most apt and those which have the best solutions with respect to the problem posed. There are two types of selections, a selection by roulette and a selection by tournament (Codio, 1995).

Once the parents have been selected, a cross is made between them, giving rise to two offspring which acquire genes from both parents. These offspring then reach the mutation stage, where one or more genes change their value in a random way (Moujahid, Inza, and Larranaga, 2008). Figure 4.3 shows the basic structure of a genetic algorithm.

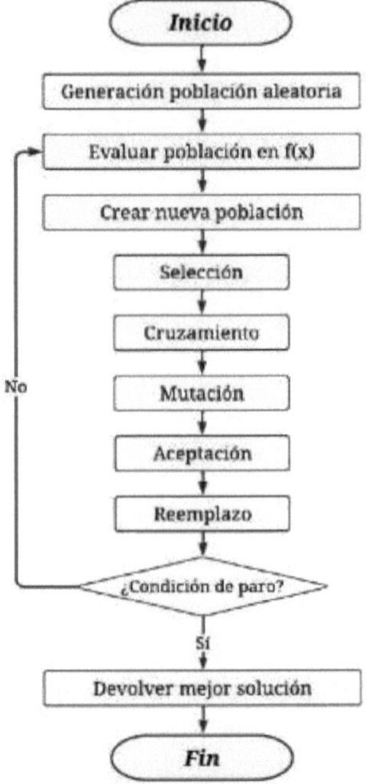

Figura 3: : Flowchart of the basic structure of a genetic algorithm
Source: Adapted from Moya Rodríguez and Méndez (2013).

TLBO (Teaching and Learning Based Optimisation) Algorithm

The teaching and learning process-based optimisation (TLBO) algorithm was proposed by Rao (2011), which seeks to simulate the behaviour experienced in a classroom, where the group of students is considered as the population, and the average grade of each student is similar to the "fitness" value of the optimisation problem posed (Rao, Kalyankar, & Waghmare, 2014).

In particular, the quality of teachers affects students' academic performance. Clearly, a good teacher will train students to get better grades (Rao, Savsani, & Vakharia, 2011). As stated earlier, these grades can be seen as quality of solutions, which are gradually improved by simulating the teacher's teaching process and interaction among students. Thus, it is shown that TLBO is an algorithm that consists of two phases, the teacher phase and the student phase.

I. **Teacher phase**

Given a set of students, i.e. an initial population, each one is evaluated and the best in terms of solution is chosen, which may be the one that yields the lowest result in the objective function as it is a minimisation problem, this best solution will be considered as the teacher, from this the teaching phase begins, where the teacher taking into account a teaching factor (TF) transmits the knowledge to the students. This factor is a random integer and is a general weighting value for the teacher's ability to teach (Cruz, Redondo, Alvarez, Berenguel, and Ortigosa, 2016), so that the teacher has the diversity to teach the student very well so that the student is completely like him or her, or otherwise that he or she manages to transmit at least one of his or her knowledge to the student. Finally, only those students who after the teaching are evaluated and improved their quality of solution will be able to pass to the next phase, otherwise they will pass as they were before this phase was applied.

IL **Student phase**

In this phase, teacher-learned students interact with each other in order to acquire more knowledge, in this case from their best peers, so a random student is grouped with another student, and he learns something new only if his peer has better grades than him (Rao et al., 2011). Figure 4.4 shows the basic structure of a TLBO algorithm.

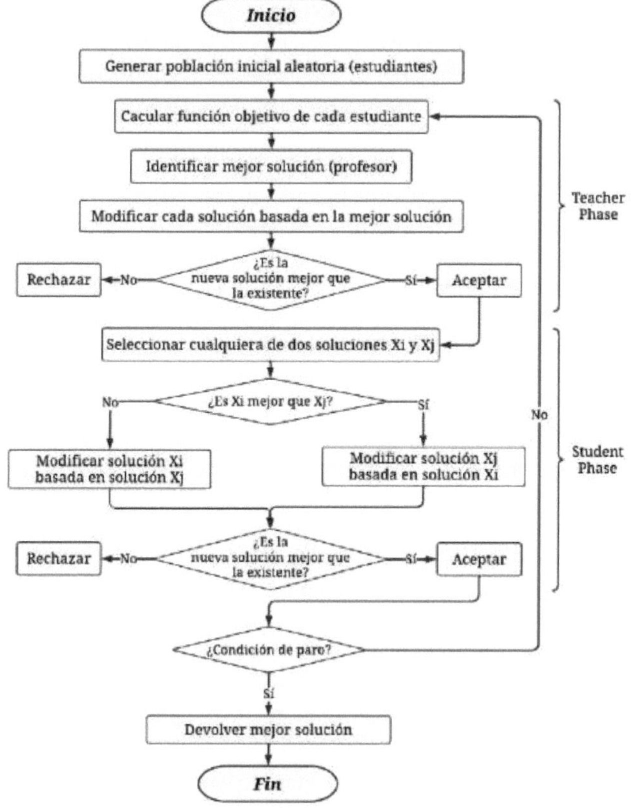

Figura 4: Flowchart of the basic structure of the TLBO algorithm.
Source: Adapted from Rao et al. (2011)

4.3. Literature review

Based on what is described in the problem statement, it is necessary to study in more depth what has been worked on in recent years regarding the integrated problem of batch sizing and production sequencing in parallel multi-machine production environments with sequence-dependent *setups*. However, some fundamental studies from much later years cannot be discarded, as they are the basis for developing and carrying out this work, such as the work done by Drexl and Kimms (1997), where their main objective was to make a literature review of the model that provides a solution to the capacitated lot sizing problem (CLSP) and the different variants that it may have, as a result of this review and given some exact solution methods, it is possible to plan approximate solution methods that meet certain conditions.

The search for solutions to batch sizing and sequencing problems has become a challenge for many researchers in recent years. Toledo, da Silva Arantes, Hossomi, França, and Akartunah (2015) managed to combine two heuristics such as *Relax and Fix* (RF) with *Fix and Optimize* (FO) to obtain a simple but efficient heuristic, called RFFO, in which RF is used to build an initial solution that FO further improves in the available computational time, this research yielded very positive results on the two-stage glass container production scheduling problem (TGCPSP), which is motivating, as the main focus of this research is directed towards approximate solution methods (Heuristics and metaheuristics).

Continuing with the search for solutions to the CLSP base problem, Almada-Lobo and James (2010) were interested in search metaheuristics for capacitated batch sizing with sequence-dependent configurations, the purpose is to develop two meta-heuristics, Variable Environment Search (VNS) and Tabu Search (TS), for the CLSP problem with the objective of determining which of the two is more efficient for this problem, with the application of these two metaheuristics, the result is that VNS outperforms TS as long as there is enough time available, i.e., which would lead to TS being more efficient in computational time.

Boonmee and Sethanan (2016) developed a metaheuristic (GLNPSO) for the multilevel batch scheduling and sizing problem in chicken egg production, applying particle swarm optimisation (PSO) and the variant (GLNPSO) with the local search procedure and the development of reinitialisation and reordering strategies to improve the solutions, From this, the GLNPSO algorithm compared to the traditional PSO shows significant improvements, including how efficient the algorithm is with respect to computational time.

In turn Parsopoulos, Konstantaras, and Skouri (2015), sought the application of an algorithm to a test set employed in previous studies, where their main objective was to thoroughly investigate the performance of the popular population-based, differential evolution (DE) algorithm for the single-item dynamic lot-sizing problem with returns and remanufacturing, resulting that DE is very competitive and can be considered a promising alternative for solving the problems considered.

Following the dynamics of the problems, Fandel and Stammen-Hegene (2006) focused their research on simultaneous batch sizing and scheduling for multi-level multi-product production through an exact model, aiming to minimise the sum of sequence-dependent setup costs, inventory costs, production costs and costs of maintaining machine setup conditions, although due to the complexity of the model, only problems with a small number of products and macro-periods can be solved optimally.

Another accurate model addressed in recent years was given by Kis and Kovács (2013), to answer a two-level uncapacitated batch size problem with backlogs, the novelty of such an approach lies in modelling the follower optimality conditions by using a primary and a dual formulation for the same problem connected by a single equation, and thus we can avoid the use of extra binary variables to model complementarity conditions which is a standard technique in bilevel optimisation.

In addition to the literature review set out in Drexl and Kimms (1997), a further review was undertaken, this time with a look at batch sizing and scheduling for perishable food products, by Chen, Berretta, Clark, and Moscato (2019), in which the articles were classified according to the characteristics of batch sizing and scheduling included in their models, and the strategies used to model perishability, something important to highlight is that the use of shelf-life constraints is the main modelling strategy used in most of the articles in this review.

Likewise, a review of the state of the art for multi-objective metaheuristics for discrete optimization problems (MODOP), was also included as part of the literature studied for this research, Liu, Li, Liu, and Guo (2020) in company with other researchers made a literature review from four perspectives, including existing multi-objective metaheuristics for MODOP, MODOP application areas, performance metrics, and test instances, from this according to the multi-objective selection mechanisms, objective metaheuristics in the literature can be classified into four categories, including dominance-based, decomposition-based, indicator-based, and hybrid selection mechanism-based.

Mor, Mosheiov, and Shapira (2020) made a study for single machine batch scheduling to minimize the (weighted) number of late orders, within this work they studied batch scheduling problems with order splitting and non-splitting, furthermore they proposed a solution obtained in polynomial time to minimize the number of late orders *(split),* providing a pseudopolynomial solution for the weighted *(split)* version, with this a type of *NP-HARD* problem for the minimum number of late orders without splitting (as an extension of the well-known container packing problem) was demonstrated, furthermore a heuristic for the no-splitting problem was introduced, consequently it is worth noting that the DP efficiently solves large instances.

To further highlight research based on metaheuristics, Pidre, Dorado, and Lorenzo (2002) addressed the problem of assigning classrooms for exams in a university centre through the application of genetic algorithm techniques, this problem takes into account a total of 5 constraints, which can be classified as spatial, temporal and human resources. The authors propose a traditional genetic algorithm, explaining that from an initial population of P members, new P members arise through crossover mechanisms, mutation and the survival or cost function, then the new population is updated and another iteration is performed. In their coding they opted for a two-dimensional vector M*J where J represents the days of other activities and M is the set of subjects that require space for exams. Among the results obtained, the best values for the probabilities were 75 % for the crossover and 2 % for the mutation, concluding that these problems can be efficiently solved by heuristic methods.

Later, Toro Ocampo Eliana M (2005) propose a hybrid method between the Chu-Beasley genetic algorithm and the Simulated Annealing (RS) algorithm for the solution of the Generalised Assignment Problem (GAP), where they create the first population using a constructive heuristic method based on sensitivity factors, and use a traditional mathematical model of the general assignment problem modified by the inclusion of penalty factors, in order to improve feasibility and reduce the computational effort. One strategy of this research to solve the PAG is to address the mutation stage of the genetic algorithm through the RS; in the crossover stage and unlike the traditional method of the genetic algorithm, they combine the two parents in such a way that only one child is born, this being an efficient local search strategy, where better answers than those obtained by Chu-Beasley were obtained.

Another case where a hybrid method was used is that of Hernández (2019). This time used for the *buffer* allocation problem that minimises the in-process inventory in open serial production lines. The metaheuristic techniques used for this approach were Genetic Algorithm (GA) and Simulated Annealing (SA). Lately, these approaches have allowed achieving satisfactory results in the study of other problems classified as *NP-HARD*. This research proposed the RS for the trajectory-based method and the GA for the population-based one, where they highlight that for the crossover phase, parent 1 is selected through a deterministic tournament, while parent 2 is chosen randomly, in order to increase the exploration of new regions of the search space. The result of the calibration performed in this work for the algorithm parameters was 60 % for the crossover probability and 13 % for the mutation probability. At the end of the study it can be noted that the proposed hybrid algorithm is more efficient than the standard RS and AG algorithms, in addition to the fact that it starts to improve its performance when analysing larger instances.

Dokeroglu (2015) developed hybrid algorithms based on *teaching-learning* optimisation (TLBO) technique for quadratic assignment problem, as well explained the TLBO based algorithm consists of two phases where all individuals are trained by a teacher in the first phase and interact with classmates to improve their knowledge level in the second phase. In this research individuals are trained with recombination operators, which do not require any parameter settings according to the original concept, and then a Robust Tabu Search (RTS) engine processes them. Experiments of the proposed hybrid sequential and parallel TLBO-RTS algorithms yield competitive performance when compared to other algorithms in the literature, these algorithms work well with random configurations, where the robustness of the proposed algorithms can be improved by increasing processors.

On the other hand, Kumar, Mittal, Soni, and Joshi (2018) focus their research on the simultaneous selection and scheduling of interdependent projects, for this they apply 3 metaheuristics which are, TLBO, Tabu Search (TS) and the hybrid TLBO-TS. In the development of this problem, two types of interdependencies were considered, mutual exclusivity and complementarity, the object of the problem being to minimise the total expected time of the selected portfolio of projects. For the TLBO and hybrid TLBO-TS algorithms, the Taguchi method was used to determine the optimal parameter levels. The algorithms were run for 100 iterations, observing that the TLBO algorithm is superior to the TS algorithm regardless of problem size and complexity, however, the hybrid TLBO-TS outperforms the other algorithms in all 4 types of problems posed, where the author concludes that "the strong local search capability of the TS helps the hybrid algorithm to reach the best result faster. Therefore, the proposed hybrid TLBO-TS algorithm not only produces better quality solutions, but also produces them faster".

Based on the above, table 4.1 presents a summary of the articles reviewed which are relevant to the development of this project, detailing in the case of approximate methods their class and type, which can be inspired by nature (N) or not based on emulation, i.e. artificial (A).

Table 4.1: Summary literature review

SUMMARY LITERATURE REVIEW					
			Algorithm type		
Authors	**Problem**	**Solution**	**N**	**A**	**Type of**

		method			algorithm
Drexl and Kimms (1997)	Batch sizing and scheduling	Exactly		XA	XA
Pidre et al (2002).	Classroom allocation	Mctahcuristics	X		GA
Toro Ocampo Elia na M (2005)	Generalised Assignment Problem (GAP)	Mctahcuristics	X		GA & RS
Fandel and Stammen-Hegene (2006)	Simultaneous batch scheduling and sizing for the production of multiple levels of multiple products	Exactly		XA	XA
Almada-Lobo and James (2010)	Capacitated batch sizing with sequence-dependent configurations	Mctahcuristics		X	VXS & TS
Kis and Kovács (2013)	Dimensioning of two-level lots	Exactly		XA	XA
Toledo et al. (2015)	Sizing of multi-level lots	Heuristics		X	RE & FO

SUMMARY LITERATURE REVIEW

Authors	Problem	Solution method	Algorithm type N	A	Type of algorithm
Parsopoulos et al (2015).	Dynamic sizing of single-item batches with returns and remanufacturing	Mctahcuristics	X		DE
Dokeroglu (2015)	Quadratic allocation	Mctahcuristics	X		TLBO
Boonmee and Sethanan (2016)	Scheduling and sizing of multi-level capacity flocks in the poultry industry	Mctahcuristics	X		PSO & GLXPSO
Kumar et al. (2018)	Integrated project selection and programming	Mctahcuristics	X		TLBO & TS
Chen et al. (2019)	Batch sizing and scheduling for perishable foodstuffs	N/A		XA	XA
Hernández (2019)	Buffer allocation that minimises in-process inventory on open serial production lines	Mctahcuristics	X		GA & RS
Liu et al. (2020)	Discrete optimisation problems	XA		XA	XA

Mor et al. (2020)	Batch programming on a single machine	Exactly	XA	XA

Source: Own elaboration

Based on the literature review made on the integrated problem of batch sizing and production sequencing in parallel multi-machine production environments with sequence-dependent *setups*, it can be observed that 27% of the investigated works are part of the exact methods presented in the previous section, 6% of them were a general literature review of the problem. However, 67% of the remaining works were developed by approximate solution methods where 77% of those works were executed based on metaheuristics, which nowadays are being very relevant in the research field. These approximate solution methods can be increasingly adapted to different problems presented in the industry, as is the case of this work, which aims to adapt different metaheuristics to the problem posed.

It is worth noting that within the genetic algorithms reviewed, it is in the work of Hernández (2019) where they make use of a design of experiments with the purpose of calibrating the parameters of the proposed algorithm, to then apply these parameters to the different case studies of the research. This statistical analysis tool is used to make an efficient exploration of the solution space, and thus obtain the variables that best adapt to the proposed problem.

From the research, it could be observed that within the literature reviewed, the proposed metaheuristics do not focus on the problem posed, however, there are very significant approaches that help the progress of this research, since some of them develop separately the parts of the problem, i.e., some focus only on the allocation and sequencing while others only on the sizing problem. In addition to this, in the literature consulted no works were found that apply the TLBO algorithm to the integrated problem of batch sizing and sequencing of production in parallel multi-machine production environments with sequence-dependent *setups*, being the application of this one of the differentials of the present work.

CHAPTER 5

METHODOLOGY

The interest of this research focuses on identifying and adapting different solution algorithms for the integrated problem of batch sizing and sequencing of production in production environments when there are multiple machines that can work in parallel and with a dependence of enlistment times with the sequence of production on the machines when changing products on each machine, which is commonly known as sequence-dependent *setups*, having as data to optimize the costs of: inventory, backlog and *setups,* in this work the production cost will not be taken into account, in order to avoid that the algorithm delays the whole production and does not produce.

This work has a quantitative approach and is classified as experimental research, as certain variables are intentionally manipulated in order to observe the effects they produce on the objective function (Sampieri, 2018).

The main focus of this work is the development of hybrid solution methods for the PMSD-LSP, where different approximate solution algorithms (Metaheuristics) and an exact model (CLSP) were adapted. This method is hybrid as it seeks that the metaheuristics provide a solution to the sequencing problem while the exact model is in charge of the sizing problem, in this way the exact model takes as input parameters the allocation proposed by the metaheuristics, the sequencing generated, in order to make the decision of how much to produce in the planning horizon, trying to minimise the aforementioned costs. Figure 5.1 illustrates the phases of this method.

The design of the algorithms proposed was carried out in such a way that they are sufficiently robust and can be adapted to any problem of this type, in addition to the algorithm being efficient in terms of computational time without neglecting the quality of response, in order to then compare the different algorithms designed between them and in relation to the exact CLSD model.

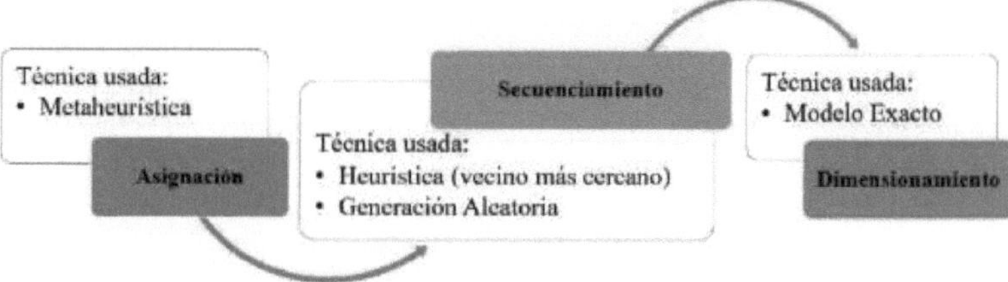

Figura 1: Phases of hybrid methods.

Source: Own elaboration

Two basic algorithms were chosen to develop the hybrid solution methods, the genetic algorithm (GA) and the teaching and learning based optimisation (TLBO) algorithm. The first algorithm was chosen due to its wide use in the field of optimisation and especially its application in the field of batch sizing and sequencing, while the use of the TLBO algorithm is justified as an adaptation and its incursion into this type of problem, seen as a differential of this research.

Once the results were obtained according to the different cases proposed, we proceeded with a descriptive analysis in order to compare whether there is a significant difference between the results obtained by each algorithm, taking into account the quality of the solutions and the computation time.

In order to solve the research problem of this work, the following method described in this section is proposed. It consists of 6 phases which are distributed throughout this research, these stages will be used as shown in Figure 5.2.

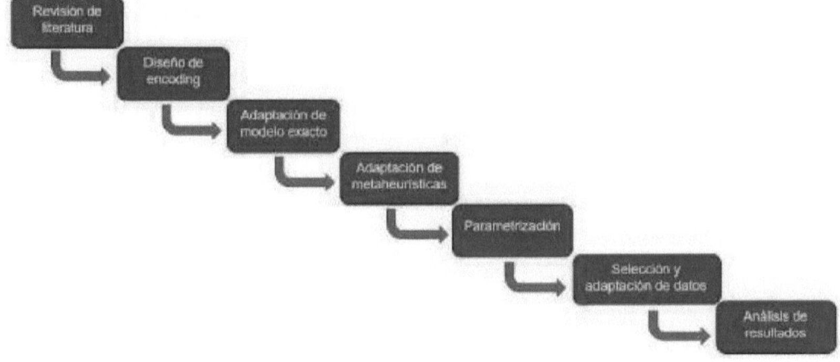

Figura 2: Phases of the investigation.
Source: Own elaboration

Based on the figure presented above, each of the phases will continue to be detailed in the following sections, emphasising and detailing how each of the phases was developed in this work.

2.1. *Encoding* design

The *encoding* developed for the execution of the different algorithms designed is a set of three 4-dimensional hypermatrices $[i * m * t * n]$, one for the assignment, one for the sequencing and one for the *setup* times, where i represents the number of products, m the number of machines, t the number of periods and n the number of solutions. The hypermatrix for the allocation has a binary encoding, with 1 representing that the product was allocated on a certain machine and period, and 0 representing the absence of this. This design can be best illustrated in Figure 5.3, where an example of the allocation generated for periods 1 and 2 in solution 1 can be seen.

[, , 1,1] → PERÍODO 1, SOLUCIÓN 1

PRODUCTOS	MÁQUINAS 1	2	3
1	1	1	0
2	0	0	0
3	0	1	1
4	0	0	1
5	1	1	0

[, , 2,1] → PERÍODO 2, SOLUCIÓN 1

PRODUCTOS	MÁQUINAS 1	2	3
1	0	0	1
2	1	0	0
3	1	0	1
4	0	0	0
5	1	1	1

Figura 3: Example of *Encoding* Assignment

Source: Own elaboration

To *encode* the sequencing and subsequently obtain the *setup* times, a permutation is generated from the elements that are 1 in the assignment *encoding*. The total of values different from 0 that the hypermatrix has for these cases will be the total number of products assigned in that machine and period, where the value generated represents the order in which the product will be produced. The *encoding* associated with the sequencing is shown in Figure 5.4, where the sequence of the assigned products is generated from the previous example.

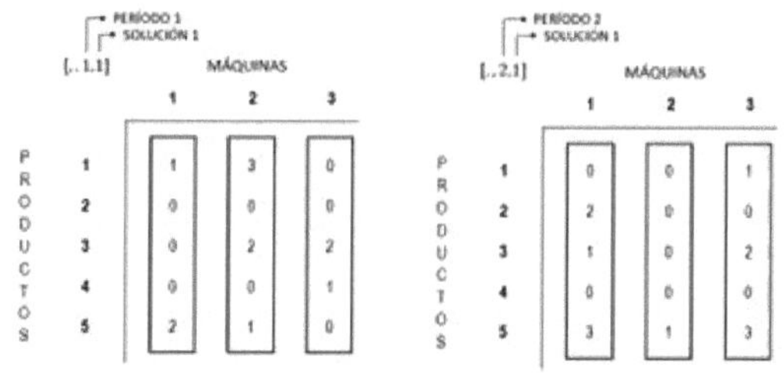

Figura 4: Sequencing *Encoding* Example

Source: Own elaboration

5.2. Adaptation of the exact model

Subsequently, we continued with the adaptation of the model, *The capacitated lot sizing problem* **CLSP,** this model is part of the main structure of the hybrid algorithms and represents the exact part, to provide an answer to the integrated problem of lot sizing and production sequencing in parallel multi-machine production environments with sequence-dependent *setups*.

The CLSP problem considers long planning periods in which multiple products can be manufactured, the problem seeks to determine how much to produce in each period in such a way that orders with lead times before the end of that period are fulfilled (Drexl and Kimms, 1997).

This model was adapted from the literature, making some modifications so that it can be adjusted to the problem of this work, another set index was added, as well as other parameters, some of these parameters are those obtained by the metaheuristic that will be detailed in the next section, these parameters are; the allocation and the sequencing. This model seeks to minimise the direct production costs, which are: inventory cost, backlog cost and *setup* cost.

The CLSP model adapted and implemented in the hybrid method developed in this work is as follows:

***I.* indices**

1, j := index for products

m := index for machines

t := index for the periods

***II.* Sets**

P := Set of products i, i $\in$ P

M: = Set of machines m, m $\in$ M

T := Set of periods t, t T ∈

***III.* Parameters**

N -.= Very large number

$C_{;tm}$:= Capacity of machine m in period t, t ∈ T, m ∈ M

$d_{t;i}$:= Demand for product i in period t, t ∈ T, i ∈ P

$prm_{;i}$ ■= Production speed of product i on machine m, m ∈ M, i ∈ P tpmi :˭ Production time of product i on machine m, m ∈ M, i ∈ P I_{Ci} := Initial inventory of product i, i ∈ P

hi := Cost of inventorying product i, i P ∈

bi := Cost of delaying product i, i ∈ P

S_m := *Setup* cost per unit time on machine m, m ∈ M

Parameters obtained by the metaheuristic

$yt_{,m,i}$ '■= 1, if a product i is assigned to a machine m in a period t. t ∈ T, m 2 M, i ∈ P

z $m_{t;;i}$ = Changeover time of a product i on a machine m in a period t. t ∈ T, m ∈ M i ∈ P

***IV.* Decision variables**

$x_{,tm,i}$ ■= Quantity of units to be produced of product i on machine m in period t. t ∈ T, m 2 M, i ∈ P

$I+$:= Quantity of units to be inventoried of a product i in a period t t ∈ T, i P ∈

Ip := Quantity of units to be backlogged of a product i in a period t t ∈ T, i ∈ P

***V.* Target function**

Min *CTotal* :

$$\sum_{t=1}^{t}\sum_{i=1}^{i} h_i \cdot I^+_{t,i} + \sum_{t=1}^{t}\sum_{i=1}^{i} b_i \cdot I^-_{t,i} + \sum_{t=1}^{t}\sum_{m=1}^{m}\sum_{i=1}^{i} S_m \cdot z_{t,m,i} \quad (5.1)$$

***VI.* Restrictions**

VII. Inventory balance restriction *m*

$$I_{0i} + \sum_{m=1}^{m} x_{t,m,i} + I^-_{t,i} = d_{t,i} + I^+_{t,i} \quad \forall t \in \mathcal{T} : t = 1, i \in \mathcal{P} \quad (5.2)$$

$$\sum_{m=1}^{m} x_{t,m,i} + I^+_{(t-1),i} + I^-_{t,i} = d_{t,i} + I^+_{t,i} + I^-_{(t-1),i} \quad \forall t \in \mathcal{T} : t > 1, i \in \mathcal{P} \quad (5.3)$$

VIII. SetUp constraint (with variable)

$$x_{t,m,i} \leq \left(\frac{C_{t,m}}{tp_{m,i}}\right) \cdot y_{t,m,i} \quad \forall t \in \mathcal{T}, m \in \mathcal{M}, i \in \mathcal{P} \quad (5.4)$$

IX. Machine capacity restriction

$$\sum_{i=1}^{i}(tp_{m,i} \cdot x_{t,m,i}) + \sum_{i=1}^{i}(z_{t,m,i} \cdot y_{t,m,i}) \leq C_{t,m} \quad \forall t \in \mathcal{T}, m \in \mathcal{M} \quad (5.5)$$

X. Non-negativity constraint

$$x_{t,m,i} \geq 0 \quad (5.6)$$

$$I^{+}_{t,i} \geq 0 \quad (5.7)$$

$$I^{-}_{t,i} \geq 0 \quad (5.8)$$

The objective function (5.1) minimises the inventory, backlog and switching costs. Equations (5.2) and (5.3) represent the standard inventory constraints. Equation (5.4) refers to the maximum amount of products that can be made if it is assigned to a machine. Equation (5.5) takes into account the capacity consumed for the production of the products and the total *setup* times. Finally, constraints (5.6), (5.7) and (5.8) represent the non-negativity conditions.

5.3. Adaptation of algorithms

5.3.1. Genetic Algorithm (GA) based methods

As its name indicates, this algorithm alludes to the genetic process that living beings carry out at the moment of reproduction, starting from an initial population of individuals, where each one represents a possible solution to a particular problem (Moujahid and cois., 2008). These individuals can be viewed as a set of chromosomes that in turn contain genes as shown in Figure 5.5.

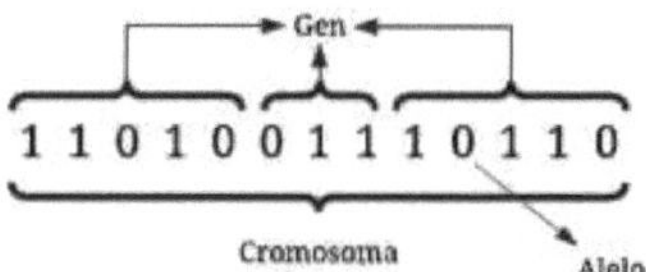

Figure 5.5: Binary genetic individual
Source: Codio (1995)

The genetic algorithm used in this work is a simple genetic algorithm which consists of the following stages:

- **Selection stage:** In this work we mainly opted for tournament selection, where the main idea is to compare between the aspiring parents and finally choose the ones with the best fit. Additionally, tournament selection was combined with roulette selection, i.e. random selection, as a modified version of the algorithm.
- **Crossover stage:** In this stage the crossover can be given in different ways, the crossover based on a point which was the operator selected to develop this work, which consists of choosing a random number between 1 and the total number of genes of each parent minus 1, in order to transmit the set of genes found in both parts of that point to the children (Diego-Más, Ballester, and Santamarina Siurana, 2020), originating two totally different new chromosomes, this process can be seen in the following way:

In addition to the point-based crossover operator, there are multipoint crossover, segmented crossover, uniform crossover and other operators that have emerged according to different application problems (Serna and Marín, 2009).

It is important to know that all selected parents will depend on a probability as to whether they can be crossed or not, each pair of parents is assigned a random number which is then evaluated with a probability usually between 0.5 and 1.0 (Moujahid and cois., 2008)In the event that this crossover probability (CP) is not exceeded, the offspring will become the same parents.

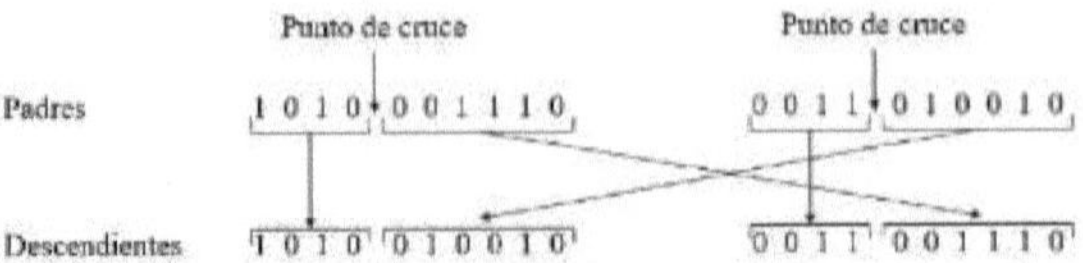

Figure 5.6: Point-based crossover operator

Source: Moujahid et al (2008).

■ **Mutation stage:** This operator is applied to each child separately, for binary encodings (as is the case in this research) it simply consists of altering a 0 to a 1 or vice versa. As in the previous stage, in this one a mutation probability (MP) is also determined and it works in the same way, a random number is assigned to a child and it will mutate if it reaches that probability, otherwise it will remain the same as it descended after the crossover. Figure 5.7 illustrates the handling of this step.

Mutated gene

Descendant1 0 1 0 0 0 1 0 0 0 0 1 0 0

Descendant Mutado 1010110010

Figure 5.7: Mutation operator

Source: Moujahid et al. (2008).

Consequently, the set of finished children is evaluated, in order to know if there was an improvement in the solutions, this new set will become part of the new population, so that the algorithm is run again in its entirety, this process is known as iteration. The number of iterations, as well as the parameters necessary to carry out the genetic algorithm are defined in section 6.2.

4 versions of the genetic algorithm were generated in this work. Two of them are based on the classical version with tournament at each iteration, one with random sequencing and one with sequencing based on the nearest neighbour heuristic. The other two versions are based on a roulette selection of parents from the second iteration, the first one with random sequencing and the second one with sequencing generated with the nearest neighbour heuristic.

The classical genetic algorithm with tournament in each population is detailed in section 5.3.1.1, while the extra genetic algorithm with tournament only in the initial population is shown in section 5.3.1.2.

5.3.1.1. Classical Genetic Algorithm

In this modality, the genetic algorithm is developed with the selection-turning process performed in each of the iterations as it is commonly known, in addition to this, two methods are applied when generating the sequencing, a totally random generation and another generation where the heuristic technique of the nearest neighbour is used.

5.3.1.1.1. Genetic Algorithm with random sequencing

The process followed by this algorithm is shown in Figure 5.8, where the difference lies in the random generation of the sequencing.

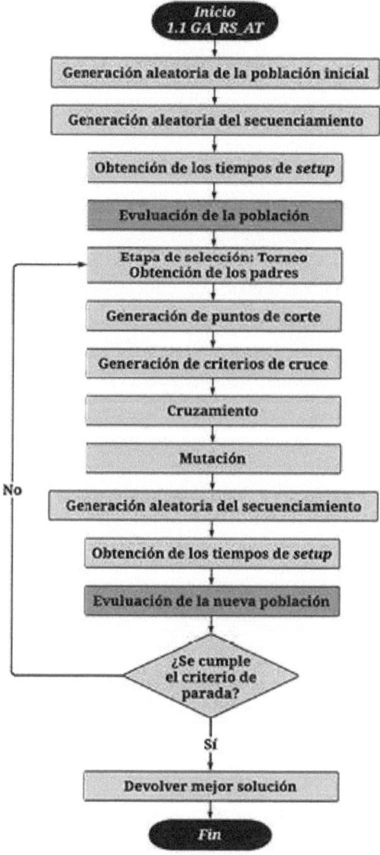

Figure 5.8: Flowchart Classical Genetic Algorithm with random sequencing
Source: Own elaboration

5.3.1.1.2. Genetic Algorithm with heuristic sequencing

The process that follows this algorithm is shown in Figure 5.9, this algorithm unlike the previous one uses the nearest neighbour heuristic for the generation of the sequencing, this heuristic is commonly known to solve the Travelling Agent Problem where it is simulated that a salesperson must visit all its customers, starting from a starting point and ending at the same point, in such a way that it minimises the different paths when visiting each of its customers (.Johnson and McGeoch, 2018).

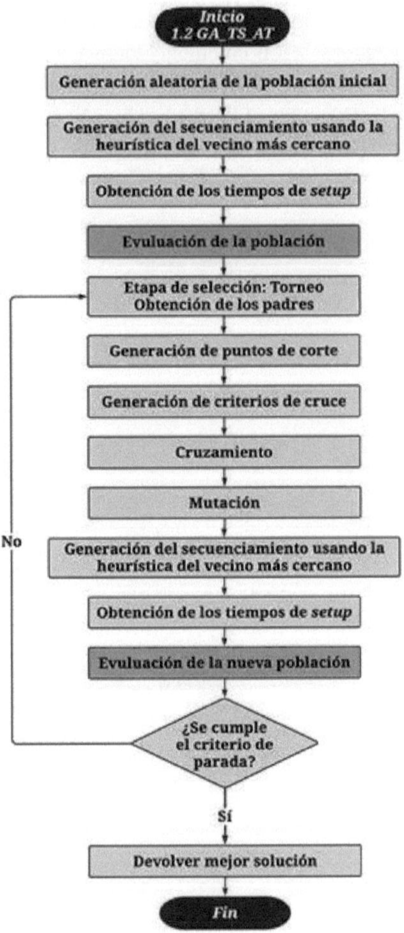

Figure 5.09: Flowchart Classical Genetic Algorithm with heuristic sequencing
Source: Own elaboration

5.3.1.2. Modified Genetic Algorithm

This is a variation of the genetic algorithm where the tournament phase is performed only once at the beginning of the tournament, i.e. with the initial population, for the other iterations the selection of parents is done by roulette, in addition to this and as was done with the classical algorithm, this reformed algorithm is divided into two, one where a completely random generation is used for sequencing and another where the generation arises using the nearest neighbour heuristic.

5.3.1.2.1. Genetic Algorithm with random sequencing

The step-by-step of this reformed algorithm is shown in Figure 5.10, in this algorithm the sequencing is randomly generated.

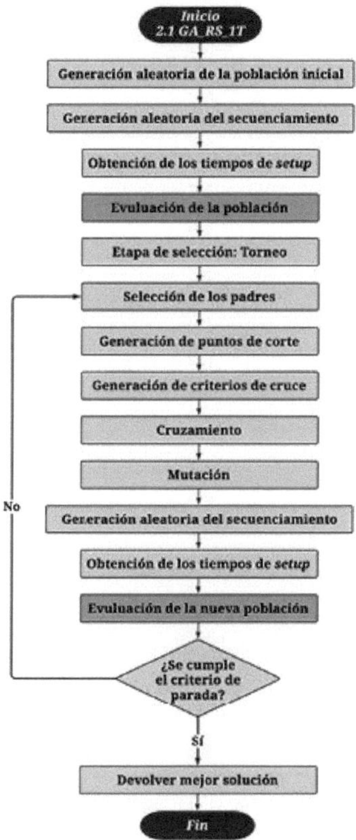

Figure 5.10: Flowchart Modified Genetic Algorithm with random sequencing

Source: Own elaboration

5.3.1.2.2. Genetic Algorithm with heuristic sequencing

The development of this reformed algorithm is shown below, where the distinctive feature is in the generation of the sequencing, as this part makes use of the heuristics explained above.

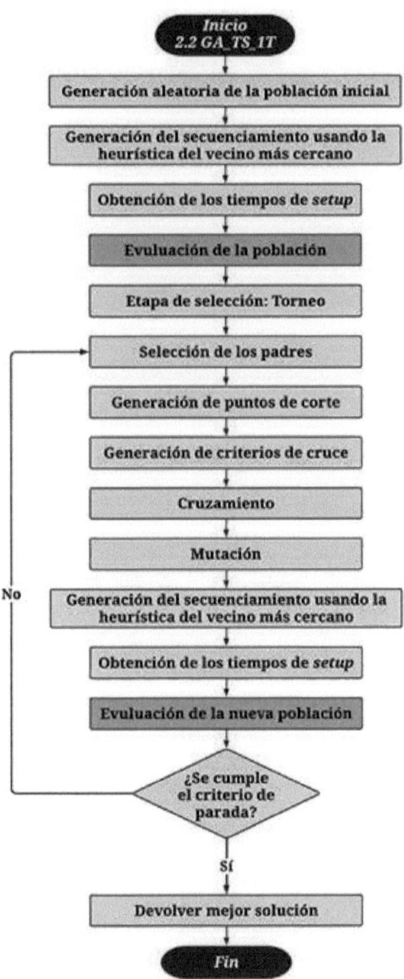

Figure 5.11: Flowchart Modified Genetic Algorithm with heuristic sequencing
Source: Own elaboration

5.3.2. Methods based on TLBO Algorithm

The TLBO is an algorithm inspired by knowledge diffusion in a classroom where students first acquire knowledge from teachers and then from classmates (Crepinsek, Liu, & Mcrnik, 2012).. The key features of the algorithm are its performance and the lack of specific search parameters, as only the population size and the number of cycles need to be specified as needed (Cruz et al., 2016).

1. **Teacher phase**

The application of this algorithm sought to support more diversification when implementing the teacher phase, as students can learn different parts of the knowledge conveyed by the teacher. This is best explained by comparing Figure 5.12 and Figure 5.13, which are based on the *encoding* used in this work to show a fraction of the teacher's teaching to one of his students.

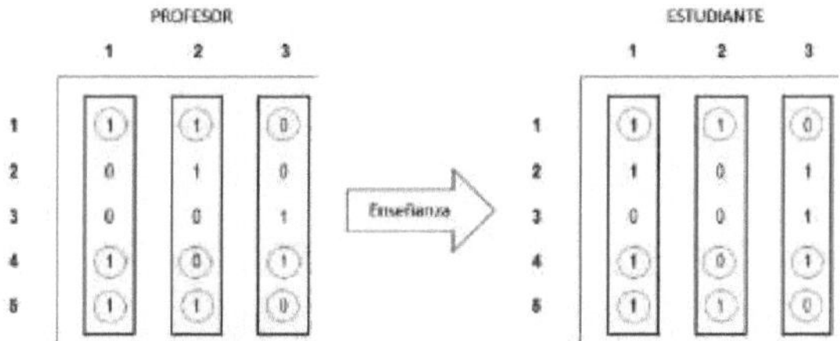

Figure 5.12: Natural education phase

Source: Own elaboration

Assuming that the number of knowledge that the teacher will teach to a given student is 3, in Figure 5.12 it can be seen that the student learnt in the same way all his parts, for the case of this work, in all his machines.

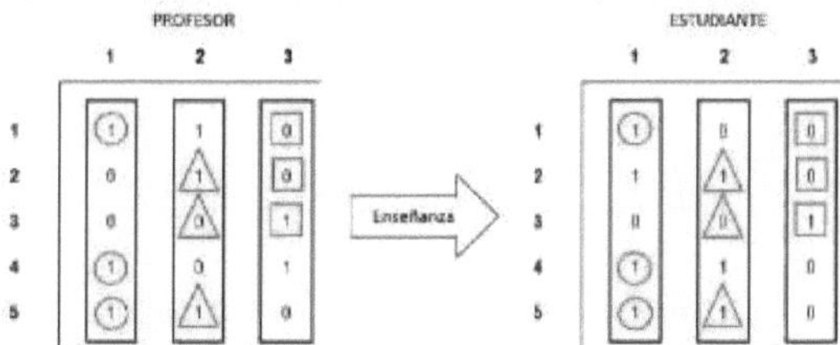

Figure 5.13: Applied teaching phase

Source: Own elaboration

Figure 5.13 illustrates the technique used in this work for the development of the teacher phase, where the number of knowledge that can be taught by the teacher is preserved.

teacher, but unlike in Figure 5.12, this knowledge varies for each of the machines. The geometric figures represent this variation, noting that the set of numbers within the figures is the same, but the position of the circles is different to that of the triangles and both different to the squares, thus influencing the diversification of the algorithm.

II. **Student phase**

As in the previous phase, in this phase, teaching factors are generated for each student teacher, thus calling the student who will teach his or her classmate, to then be re-evaluated and taken to the final population only if their solution is improved after the change (Cruz et al., 2016). This final population will be the initial population for the next iteration of the algorithm.

5.3.2.1. Random sequencing TLBO algorithm

The way this algorithm works is illustrated in figure 5.14, where the sequencing is randomly generated.

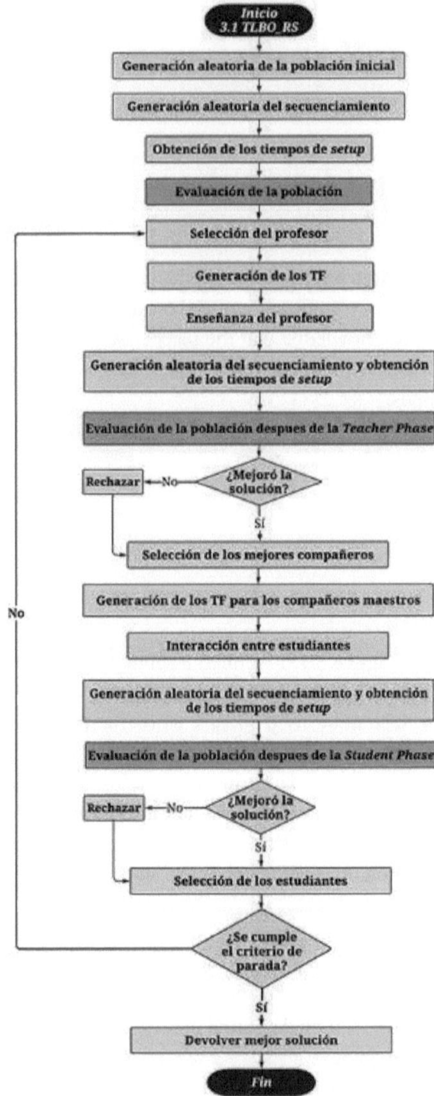

Figure 5.14: Flowchart TLBO algorithm with random sequencing
Source: Own elaboration

5.3.2.2. TLBO algorithm with heuristic sequencing

The procedure exercised by this algorithm is shown in Figure 5.15, and unlike the previous one, the sequencing is generated using the nearest neighbour heuristic.

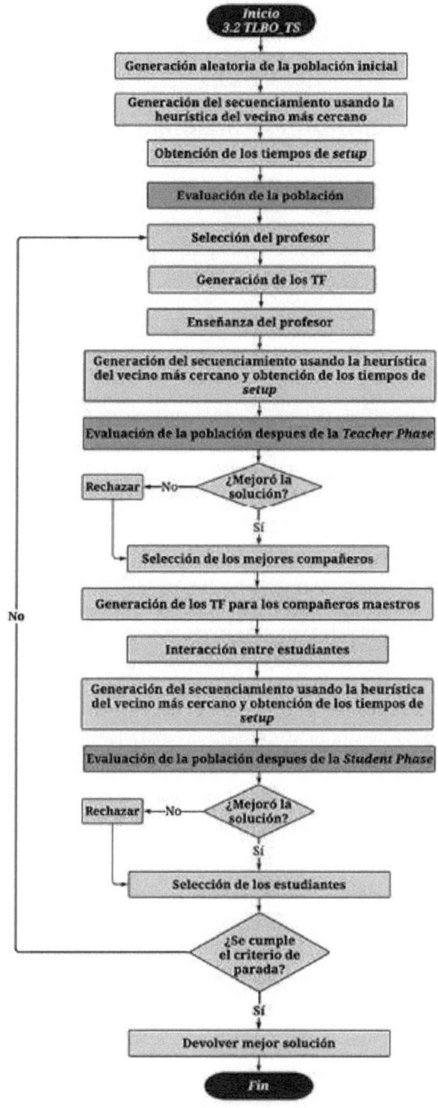

Figure 5.15: Flowchart TLBO algorithm with heuristic sequencing
Source: Own elaboration

CHAPTER 6

COMPUTATIONAL EXPERIMENTS

6.1. Selection and adaptation of instances

The data set for the application of the algorithms proposed was taken from Piñeros, Toscano, Ferreira, and Morabito (2021), taking into account that different modifications were made and only the necessary parameters were used that were adjusted to the problem posed. The original data were obtained and created for the problem of batch sizing and sequencing in the production process of fruit-based beverages.

In the production of fruit-based beverages, there are two production stages: tank preparation and packaging on the production lines. This production process has some specific characteristics, such as periodic cleaning and synchronisation between the two production stages, which makes it difficult to optimise production planning and scheduling, as they substantially increase the number of variables. These characteristics were disregarded when adapting the data, since the problem studied contemplates production in a single stage, and variables such as tank cleaning and product perishability, among others, are not taken into account.

The instances used are useful in this research because they are based on real data, illustrate a more practical view of the problem and its complexity. Furthermore, they represent different scenarios of real companies due to differences in cost and capacity parameters. The dataset consists of a collection of instances of classical lot-sizing and scheduling problems with sequence-dependent setup times and costs and changeover times satisfying the triangular inequality (Piñeros et al., 2021).

The 4 datasets proposed by the authors, each with 23 instances, were adapted.

Where groups 2, 3 and 4 are variations of group 1. These instances consist of real data and data generated from these real instances.

That said, the dataset has the following characteristics:

- **Group 1 (Gl):** Group of instances created from real data.
- **Group 2 (G2):** Retains the same data as group 1, but its capacity decreases by 10%.
- **Group 3 (G3):** The difference is in the alteration of inventory and backlog costs.
- **Group 4 (G4):** Acquires the changes made in groups 2 and 3.

The *setup* cost is the average of the matrix proposed in the original data. The standard deviation of this matrix was calculated and values of approximately 0.2% were obtained in most instances, with instance 1 being the highest at 0.8%, so it is valid to use the average and apply it to any product.

Given that in the proposed data there are two matrices for the *setup* times, one for each stage, for the development of this work the highest *setup* times were chosen, i.e. those of line 2, with this the production process was not altered, as to produce another product in a tank it is necessary for the line to be available.

From the data proposed above, a series of changes were made according to the problem developed in this work, these changes and/or adaptations are shown in table 6.1.

Parámetros	Desripción	¿Utilizacion?	Adaptación
J	Articulos	✓	Productos
M	Tanque / Línea	✓	Máquinas
T	Periodos	✓	Periodos
I_{j0}^{+} I_{j0}^{-}	Inventario/Atraso inicial	✓	Inventario inicial
te^{I} te^{II}	Tiempo de limpieza temporal en un tanque / linea	✗	
te_{max}^{I} te_{max}^{II}	Tiempo maximo transcurrido desde la ultima limpieza temporal en un tanque / linea	✗	
d_{jt}	Demanda	✓	Demanda
ρ_j	Cantidad de bebida en una unidad del item	✗	
cap_{mt}	Capacidad disponible en un tanque / linea	✓	Capacidad disponible en una máquina
b_{ij}^{I} b_{ij}^{II}	Tiempo de cambio de producción etapa 1 / etapa 2	✓	Tiempo de *Setup*
ub_j lb_j	Cantidades maximinas/minimas de produccion	✗	
pt	Tiempo de producción de bebida en un lote	✗	
S_m	Tiempo empleado por la linea en embotellar un litro de bebida	✓	Tiempo de producción
h_j^{+} h_j^{-}	Costos de inventario/atraso	✓	Costos de inventario/atraso
c_{ij}	Costo del proceso de cambio	✓	Costo de *Setup*
Ct	Costo de limpieza temporal	✗	
O_{mt}	Conjunto ordenado de lotes disponibles	✗	
Q_{mt}^{I} Q_{mt}^{II}	Conjunto ordenado de limpiezas temporales disponibles	✗	

Parameters	Description	Utilisation?	AdaptationI
	Articles	>/	Products
	Tank / Line	>/	Machines
	Periods	>/	Periods
	Inventory/Initial backlog	y/	Initial Inventory
	Temporary cleaning time in a tank / line	x	
	Maximum time elapsed since last temporary cleaning in a tank/line	x	
	Demand	y	Demand
	Quantity of beverage in a unit of the item	x	
	Available capacity in a tank / line	z	Available capacity on a machine
	Production changeover time stage 1/stage 2	Z	*Setup* time
	Maximum/minimum production quantities	x	
	Production time of beverage in a batch	x	
	Time taken by the line to bottle one lithium drink	Z	Production time
	Inventory/stocking costs	Z	Inventory/stocking costs
	Cost of the change process	Z	*Setup* cost
	Cost of temporary cleaning	x	
	Orderly set of available lots	x	
	Orderly set of temporary cleanings available	x	

Table 6.1: Data adaptation

Source: Adapted from Piñeros et al. (2021).

In addition to the 4 groups adapted from the scheduling problem in the fruit drinks production process, a group 5 was created where the following changes were made based on Grupol, the reasons for the creation of this additional group are given below:

- Since the 4 proposed groups had no initial inventory, in this group the initial inventory was generated using a random number between 0 and 15 % of the demand.
- The *setup* times vary for each of the machines, the original times were left for machine 1 and from this machine onwards the times were halved and increased by 1.5 for the following machines. In order to have a greater diversity in this data set.
- The capacity of the machines was reduced, in order to adapt a little more to real-life scenarios.

6.2. Parameterisation

Of the two types of algorithms developed (Genetic Algorithm and TLBO), according to the literature the only one that needs input parameters to work is the genetic algorithm, for this reason a design of experiments was developed in order to choose the parameters of this algorithm that best fit the problem posed.

As mentioned above, the parameters necessary for the execution of genetic algorithms are the crossover probability (CP), the mutation probability (MP), the number of solutions and the number of iterations, this last parameter is also applied in the TLBO algorithm. According to the literature reviewed, there are different models that propose different values when choosing the probabilities, in the research by Aurelio, Pa checo, Hamacher, and Almeida (2015) the following parameters are proposed, CP=0.9, MP 0.3, while in the work developed by Diego-Más and cois. (2020) states that the crossover probability is usually set in a range of 0.6 and 0.95, and the mutation probability in much lower values, likewise Goldberg (1989) states that the value of the crossover probability ranges between 0.6 and 1, usually these parameters hover around the aforementioned values. In order to increase the diversification when running the algorithms, several possible values are proposed in order to determine the parameters to be used with the design of experiments.

CP: {0.65 , 0.75 , 0.85}

MP: {0.15 , 0.25 , 0.35}

Solutions: {100 , 200 , 500}

The aforementioned values were run in the genetic algorithm with tournament only in the initial population and generation of the sequencing heuristic, as this contains the two innovative cases developed in this work, the inclusion of the nearest neighbour heuristic together with the combination of tournament and roulette for the selection stage. This algorithm was run with instances 20 and 21 of group 1, taking into account that they are instances with the largest scenarios of the selected dataset, in order to generate greater confidence when analysing the results. The design of experiments can be found in the appendix of this paper (see section 7.2).

From the design of experiments carried out, it can be concluded that the probabilities of crossover and mutation do not affect the algorithm at all in terms of both the quality of the solution and the computation time, for this reason it was decided to choose the mean value of those proposed, so that the probabilities for the genetic algorithm are: CP=0.75 and MP 0.25.

For the size of the solutions, given that they also had no impact on the variables, it is worth mentioning that the main priority of a metaheuristic is that it yields a feasible solution in the shortest possible time, in view of this, it is sought that the algorithms perform efficiently, which is why it was decided that the size of the solutions will be the minimum, in this case of size 100.

6.3. Comparison of the proposed algorithms

The 6 proposed algorithms were run on the 5 data sets of 23 instances, 4 genetic algorithms differing in sequencing generation and tournament application in the selection stage, and 2 algorithms based on teaching and learning optimisation (TLBO) also differing in sequencing generation.

The algorithms were run on a computer with the following characteristics, brand: HP ProBook 445 G8 Notebook PC, operating system: Windows 10 pro x64, RAM size: 16 GB (15.3 GB usable), processor: AMD Ryzen 7 PRO 5850u, processor speed: 1.9 GHz.

In order to summarise the name of the algorithms when comparing their results and without leaving aside the description contained in each of them, the compact titles of these metaheuristics are the following:

- GA_RS_AT: Genetic Algorithm with random sequencing generation and tournament performed in all iterations.
- GA_RS_1T: Genetic Algorithm with random sequencing generation and tournament performed on the initial population.

- GA_TS_AT: Genetic Algorithm using the nearest neighbour heuristic for sequencing and tournament generation performed in all iterations.
- GA_TS_1T: Genetic Algorithm using the nearest neighbour heuristic for sequencing and tournament generation performed on the initial population.
- TLBO_RS: TLBO algorithm with random sequencing generation.
- TLBO_TS: TLBO algorithm using the nearest neighbour heuristic for sequencing generation.

All the developed algorithms obey 3 stop criteria, which can be for reaching an estimated time limit, in this case 1 hour, for fulfilling the total number of iterations or for not improving the solution in a run of 75 iterations, this last criterion is made because in the case that the algorithm gets stuck in a local optimum, saving its computation time.

It is worth noting that all algorithms were run in R *software* version 4.1.2 for the scheduling part, i.e. for the production sequencing, for the batch sizing and evaluation part, i.e. when using the exact CLSP model, it was run using the MINOS solver, given the continuous nature of the sizing problem.

Now, in order to have more confidence in the conclusion, each of the algorithms were run 3 times and with a total of 500 iterations, the results obtained with the best solutions are shown below.

COMPARISON BETWEEN GENETIC ALGORITHMS THAT INCLUDE THE TOURNAMENT IN ALL ITERATIONS

SHOUT 1 INSTANCE	GA RS AT		GA TS AT		SCREAM 2 INSTANCE	GA RS AT		GA TS AT		SHOUT 3 INSTANCE	GA RS AT		GA TS AT		SHOUT 4 INSTANCE	GA RS AT		GA TS AT		SHOUT 5 INSTANCE	GA RS AT		GA TS AT	
	Bobj	Time	Bobj	Time		Bobj	Time	Bobj	Time		Bobj	Time	Bobj	Time		Bobj	Time	Bobj	Time		Bobj	Time	Bobj	Time
1	1,85	363	1,35	454,4	1	**1,85**	360,9	1,85	360	1	1.85	356,9	1,85	374,3	1	1,85	361	1,85	377,7	1	1.85	376,6	1,85	377,4
2	1,85	393	1,35	473,8	2	1,85	404,1	1,85	369,5	2	1,85	373,6	1,85	378,4	2	1,85	375,5	1,85	377,3	2	1,85	361,5	1,85	350,7
3	15,34	513	15,34	507	3	15,34	406	15,34	399	3	15,34	423	15,34	425	3	15,34	410	15,34	407	3	16	498	16	409
4	14,04	409	14,04	462	4	14,04	385	14,04	399	4	14,04	368	14,04	435	4	14,04	383	14,04	**416**	4	24	430	23	479
5	14	377	13,53	486	5	14	377	13,53	471,3	5	14	395	13,53	568,85	5	14	389	14	405	5	22	441	22	579
6	0	1141	62	801,5	6	5,6	1158,5	5,7	1183,5	6	12	738	13,62	1009,35	6	12	506	5,7	625,3	6	29	773	44	635
7	5	555	5	608	7	4	704	3	**564**	7	3	741	4	558	7	4	678	3	659	7	2497567	676	3165159	586
8	3	547	3,96	638,6	8	4	**664**	3	480	8	3	750	3	743	8	**0,99**	1173,45	4	861	8	11	1450	8,9	1199
9	87	824	82	728	9	95	614	90	545	9	93	665	86	638	9	91	535	82	433	9	107	503	94	1482
10	14	1003	12,55	714,93	10	9,3	**958**	11	1035	10	20	782	19	595	10	15	847	16	644	10	33	677	**46**	578
11	11	927	11,05	650,38	11	14	1010	6,05	734,15	11	11	1058	14	987	11	14	986	8,8	981,5	11	32	644	32	877
12	6,3	545,9	6,3	971,9	12	7Д	474,4	72	626,3	12	8,1	518,7	6,3	667,8	12	4,5	1251,1	6,3	745,2	12	1160213	1031	1232010	1245
13	19	924	21	939	13	22	929	14	1576	13	14	1152	22	792	13	22	664	25	991	13	29794262	294	25961909	479
14	250	1014	232,99	851,87	14	259	945	242	767	14	264	690	227	728	14	254	1216	245	574	14	483828	819	413611	1161
15	380	591	314	955	15	345	753	316	926	15	374	486	330	538	15	350	887	319	799	15	375	709	347	656
16	370	520	334,51	528,15	16	342	**464**	320	809	16	385	452	330	1014	16	379	466	314	1085	16	399	633	329	818
17	332	475	293	720	17	332	453	287	1072	17	326	570	297	784	17	329	552	280	586	17	343	638	305	677
18	817	743	690,44	718,31	18	828	561	658	**1232**	18	772	635	689	747	18	799	810	666	913	18	845	987	708	1064
19	801	893	674	852	19	834	868	717	668	19	822	1276	708	705	19	773	1081	721	975	19	901	935	791	699
20	859	532	685,63	1141,8	20	816	666	685,2	743,3	20	**804**	1234	670	607	20	831	599	700	750	20	907	918	775.85	1482,01
21	1932	685	1557	841	21	1911	769	1534	781	21	1934	897	1539	829	21	1909	1014	1538	835	21	2123	237	1787	275
22	1928	851	1600	841	22	1956	653	1575	821	22	**1950**	858	1503	1293	22	1907	871	1585	1196	22	65533847	2380	8592012	471
23	2049	948	1621	992	23	2067	1272	1625	1024	23	2073	776	1628	988	23	2113	698	1609	1101	23	**2462**	238	1802,9	265,8

Table 6.2: Results of genetic algorithms (GA) including the tournament in all iterations (TA)
Source: Own elaboration

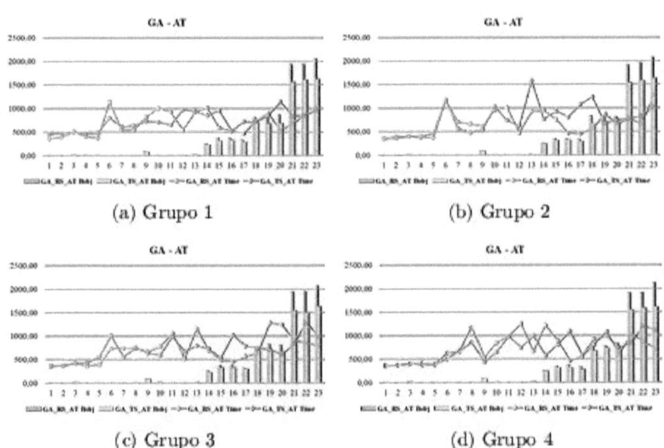

(a) Grupo 1 (b) Grupo 2

(c) Grupo 3 (d) Grupo 4

Figura 1: : Comparison of solution quality vs. computation time
Source: Own elaboration

Comparison between genetic algorithms that include the tournament in all iterations

According to table 6.2, the average time for the GA_RS_AT algorithm to find a feasible answer was 687s, this time increases by approximately 30% for large instances, while for small instances it is reduced by up to 50%. On the other hand, the average time used by the GA_TS_AT algorithm is 734s, for this algorithm the time for small instances can be reduced by up to 60 %, while for large instances the time can be doubled.

From Figure 6.1 presented above, the behaviour of the two models can be analysed, observing that the time used by the GA_TS_AT algorithm in most cases exceeds that of the GA_RS_AT algorithm, and it can also be analysed that this model achieves a better solution in a shorter time. For all groups, the superiority of the GA_TS_AT model is noted, obtaining better results. It should be noted that the difference in time of both models would be in an average of 7 %.

COMPARISON BETWEEN GENETIC ALGORITHMS THAT INCLUDE THE TOURNAMENT IN THE INITIAL POPULATION

GROUP 1	GA_ RS_1T		GA TS IT		GROUP 2	GA RS IT		GA TS IT		GROUP 3	GAJ RS_1T		GA_TS_1T		GROUP 4	GA_RS_1T		GA_TS_1T		GROUP 5	GA_RS_1T		GA_TS_1T	
INSTANCE	Bobj	Time	Bobj	Time	INSTANCE	Bobj	Time	Bobj	Time	INSTANCE	Bobj	Time	Bobj	Time	INSTANCE	Bobj	Time	Bobj	Time	INSTANCE	Bobj	Time	Bobj	Time
1	1,85	136,5	1,85	86,9	1	1,85	**122**	1,85	90,4	1	1,85	**121,7**	1,85	92,9	1	1,85	120,8	1,85	92,1	1	1,85	100,6	1,85	138,6
2	1,85	125,2	1,85	87,7	2	1,85	124,9	1,85	94,5	2	1,85	123,9	1,85	96,2	2	1,85	123,5	1,85	107,1	2	1,85	106,6	1,85	78,4
3	15,34	149	15,34	121	3	15,34	130	15,34	91	3	15,34	136	15,34	101	3	15,34	124	15,34	94	3	16	123	**16**	151
4	14,04	134	14,04	**118**	4	14,04	211	14,04	99	4	14,04	184	14,04	92	4	**14,04**	137	14,04	124	4	24	128	24	92
5	15	126	14	**111**	5	14	126	14	98	5	14	136	14	96	5	14	162	14	141	5	20	133	20	118
6	18	560	17	144	6	19	183	28	185	6	27	155	19	150	6	37	127	26	195	6	64	144	69	146
7	5,9	190,4	7,9	138,4	7	6,9	166,2	7,9	102	7	6,9	227,1	6,9	105,5	7	7,9	128,9	8,9	**112**	7	12067951	134	11670919	111
8	6,9	258,8	6,9	213,8	8	5	143	6,9	154,7	8	5,9	162,8	5	293	8	7,9	130,7	5,9	156	8	3092626	135	8267234	100
9	103	**131**	84	**116**	9	**111**	149	112	107	9	116	**131**	89	135	9	102	342	**112**	145	9	154	**151**	141	113
10	47	285	41	179	10	38	491	47	223	10	47	174	35	251	10	38	291	42	149	10	77	366	85	102
11	35	237	45	157	11	44	163	42	133	11	44	276	35	141	11	45	193	42	**93**	11	80	**161**	70	220
12	9,9	215,1	9,9	201,3	12	9,9	196,5	12	101	12	8,1	396,2	9	198	12	9,9	**177,3**	11	155	12	3171014	158	3068416	235
13	60	305	45	285	13	49	**447**	63	144	13	55	157	66	96	13	54	185	66	190	13	37570787	60	37848151	96
14	274	745	265	152	14	9364	152	231	222	14	300	163	274	185	14	283	183	622	120	14	2224287	332	1310820	146
15	369	159	310	214	15	354	285	309	168	15	372	184	315	164	15	361	156	**316**	265	15	409	239	327	165
16	391	239	331	144	16	376	362	332	228	16	351	438	292	197	16	362	207	332	**140**	16	417	223	337	152
17	340	147	291	179	17	314	232	281	**217**	17	304	193	289	124	17	284	309	304	149	17	350	225	310	143
18	808	385	722	186	18	854	228	689	386	18	883	166	748	142	18	839	328	716	**217**	18	943	257	804	259
19	851	188	785	146	19	861	539	722	156	19	837	249	759	272	19	865	324	739	228	19	954	185	839	254
20	842	549	751	197	20	870	277	726	235	**20**	838	381	713	233	20	870	**316**	730	**204**	20	962	275	839	162
21	1893	445	1570	273	21	1915	212	1573	257	21	1890	379	1551	213	21	1905	365	1496	362	21	565552472	98	9414435	276
22	1987	266	1608	198	22	1979	295	1545	329	22	1951	239	1608	262	22	1874	306	1554	176	22	2157	143	17513601	121
23	2056	548	1668	267	23	2125	275	1677	178	23	2092	**231**	1615	261	23	2118	249	**1617**	180	23	1494456211	296	824974338	90

Table 6.3: Results genetic algorithms (GA) including the tournament in the initial population (IT)

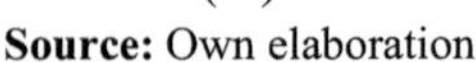

Source: Own elaboration

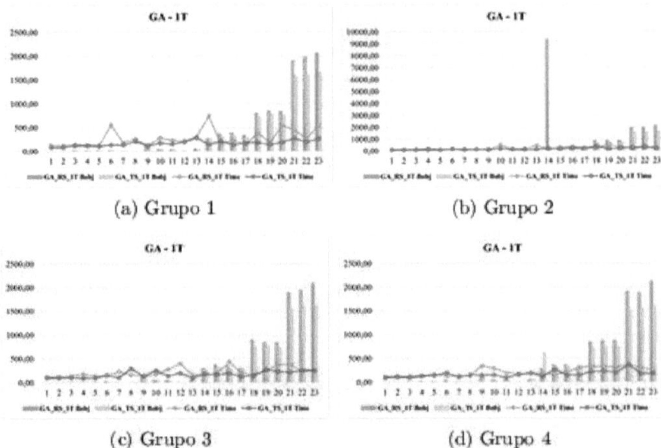

(a) Grupo 1 (b) Grupo 2

(c) Grupo 3 (d) Grupo 4

Figura 2: : Comparison of solution quality vs. computation time

Source: Own elaboration

Comparison between genetic algorithms that include the tournament only in the initial population.

For the comparison of these two genetic algorithms, it can be concluded from table 6.3 that the average time of the GA_RS_1T had an average of 227s to yield a feasible response, generating inefficient responses. On the other hand, the average time for the GA_TS_1T algorithm was 166s. The comparison of this pair of algorithms highlighted that both models are very efficient in terms of computational time. However, the quality of the responses of group 5 was far from optimal for both algorithms.

From Figure 6.2, which compares computation time vs. solution quality, it can be seen that the GA_TS_1T algorithm presents better results in terms of computation time and solution quality than the GA_RS_1T algorithm, since the latter algorithm uses more time to provide a feasible answer, and it can even be specifically noted in Figure 6.2b that the GA_RS_1T algorithm for instance 14 obtains a solution that is too far away compared to that of its opponent, however, in that same group the two algorithms have very similar times, only in instances 10, 13 and 19 is a slight difference between their times visible.

COMPARISON BETWEEN GENETIC ALGORITHMS THAT RANDOMLY GENERATE SEQUENCING

SHOUT 1 INSTANCE	GA RS AT		GA RS IT		GROUP 2 INSTANCE	GA RS AT		GA RS IT		SHOUT 3 INSTANCE	GARSAT		GA_RS_1T		GROUP 4 INSTANCE	GARSAT		GA_RS_1T		SHOUT 5 INSTANCE	GARSAT		GA_RS_1T	
	Bobj	Time	Bobj	Time		Bobj	Time	Bobj	Time		Bobj	Time	Bobj	Time		Bobj	Time	Bobj	Time		Bobj	Time	Bobj	Time
1	1,85	363	1,85	136,5	1	1,85	360,9	1,85	122	1	1,85	356,9	1,85	**121,7**	1	1,85	361	1,85	120,8	1	1,85	376,6	1,85	100,6
2	1,85	393	L85	125,2	2	1,85	404,1	1,85	124,9	2	1,85	373,6	1,85	123,9	2	1,85	375,5	1,85	123,5	2	1,85	361,5	1,85	106,6
3	15,34	513	15,34	149	3	15.34	406	15,34	130	3	15,34	423	15,34	136	3	15,34	410	15,34	124	3	16	498	16	123
4	14,04	409	14,04	134	4	14,04	385	14,04	211	4	14,04	368	14,04	184	4	14,04	383	14,04	137	4	24	430	24	128
5	14	377	15	126	5	14	377	14	126	5	14	395	14	136	5	14	389	14	162	5	22	441	20	133
6	0	1141	18	560	6	5,6	1158,5	19	183	6	12	738	27	155	6	12	506	37	127	6	29	773	64	144
7	5	555	5,9	190,4	7	4	704	6,9	166,2	7	3	741	6,9	227,1	7	4	678	7,9	128,9	7	2497567	676	12067951	134
8	3	547	6,9	258,8	8	4	664	5	143	8	3	750	5,9	162,8	8	0,99	1173,45	7,9	130,7	8	11	1450	3092626	135
9	87	824	103	131	9	95	614	111	149	9	93	665	116	131	9	91	535	102	342	9	107	503	154	151
10	14	1003	47	285	10	93	958	38	491	10	20	782	47	174	10	15	847	38	291	10	33	677	77	366
11	11	927	35	237	11	14	1010	44	163	11	11	1058	44	276	11	14	986	45	193	11	32	644	80	161
12	6,3	545,9	9,9	215,1	12	7,2	474,4	9,9	196,5	12	8,1	518,7	8,1	396,2	12	4,5	1251,1	9,9	177,3	12	1160213	1031	3171014	158
13	19	924	60	305	13	22	929	49	447	13	14	1152	55	157	13	22	664	54	185	13	29794262	294	37570787	60
14	250	1014	274	745	14	259	945	9364	152	14	264	690	300	163	14	254	1216	283	183	14	483828	819	2224287	332
15	380	591	369	159	15	345	753	354	285	15	374	486	372	184	15	350	887	361	156	15	375	709	409	239
16	370	520	391	239	16	342	464	376	362	16	385	452	351	438	16	379	466	362	207	16	399	633	417	223
17	332	475	340	147	17	332	453	314	232	17	326	570	304	193	17	329	552	284	309	17	343	638	350	225
18	817	743	808	385	18	828	561	854	228	18	772	635	883	166	18	799	810	839	328	18	845	987	943	257
19	801	893	851	188	19	834	868	861	539	19	822	1276	837	249	19	773	1081	865	324	19	901	935	954	185
20	859	532	842	549	20	816	666	870	277	20	804	1234	838	381	20	831	599	870	316	20	907	918	962	275
21	1932	685	1893	445	21	1911	769	1915	212	21	1934	897	1890	**379**	21	1909	1014	1905	365	21	2123	237	565552472	98
22	1928	851	1987	266	22	1956	653	1979	295	22	1950	858	1951	239	22	1907	871	1874	306	22	655338470	238	2157	143
23	2049	948	2056	548	23	2067	1272	2125	275	23	2073	776	2092	231	23	2113	698	2118	249	23	2462	238	L494E+09	296

Table 6.4: Results of genetic algorithms (GA) generating random sequencing (RS)
Source: Own elaboration

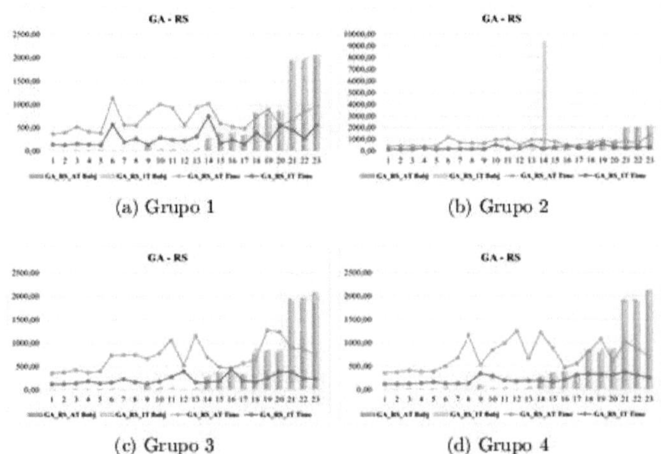

(a) Grupo 1 (b) Grupo 2 (c) Grupo 3 (d) Grupo 4

Figura 3: : Comparison of solution quality vs. computation time
Source: Own elaboration

Comparison between genetic algorithms that generate sequencing randomly

Taking into account table 6.4, this section analyses the genetic algorithm in its classical version, but with the tournament variations, the genetic version with the tournament version only in the initial population being more efficient with respect to computation time than the algorithm with the tournament applied in all iterations. The average time of the GA_RS_AT algorithm was 687s, while the average time of the GA_RS_1T algorithm was 227s, optimising the computation time for this modified version of the genetic algorithm by 33%.
With respect to the graphs presented in Figure 6.3, it can be observed that when applying the mechanism of carrying out the tournament only in the initial population, the time was always lower in all the groups in relation to the algorithm where the tournament was carried out iteratively. Additionally, the quality of the solution was also good, the results obtained by the GA_RS_1T algorithm were very similar to the results obtained by the GA_RS_AT algorithm. It is worth noting that in the tests performed there was one case where the GA_RS_1T algorithm was substantially different from the response obtained by the GA_RS_AT algorithm, but in general, the idea of only performing the tournament at the beginning of each population had good results both computationally and in terms of the quality of the solution.

COMPARISON BETWEEN GENETIC ALGORITHMS USING NEAREST NEIGHBOR HEURISTICS FOR SEQUENCING GENERATION

GROUP 1 INSTANCE	GA TS AT		GA_TS_1T		GROUP 2 INSTANCE	GA TS AT		GA TS LT		GROUP 3 INSTANCE	GA TS AT		GA_TS_1T		GROUP 4 INSTANCE	GA TS AT		GA_TS_1T		GROUP 5 INSTANCE	GA TS AT		GA_TS_1T	
	Bobj	Time	Bobj	Time		Bobj	Time	Bobj	Time		Bobj	Time	Bobj	Time		Bobj	Time	Bobj	Time		Bobj	Time	Bobj	Time
1	**1,B5**	454,4	1,85	86,9	1	1,85	360	1,85	90,4	1	1,85	374,3	1,85	92,9	1	1,85	377,7	1,85	92,1	1	1,85	377,4	1,85	138,6
2	1.85	473,8	1,85	87,7	2	1,85	369,5	1,85	94,5	2	1.85	378,4	1,85	96.2	2	1,85	377,3	1,85	107,1	2	1,85	350,7	1,85	78,4
3	15.34	507	15,34	**121**	3	15,34	399	15,34	91	3	15.34	425	15,34	101	3	15,34	407	15,34	94	3	16	409	16	151
4	14,04	462	14,04	118	4	14,04	399	14,04	99	4	14,04	435	**14,04**	92	4	14,04	416	14,04	124	4	23	479	24	92
5	13.53	486	14	**111**	5	13,53	471,3	14	98	5	13.53	568,85	14	96	5	14	405	**14**	141	5	22	579	20	118
6	6.2	801,5	17	**144**	6	5,7	1183,5	28	185	6	13.62	**1009,4**	19	150	6	5,7	625,3	26	195	6	44	635	69	146
7	5	608	7,9	138,4	7	3	564	7,9	102	7	4	558	6,9	105,5	7	3	659	8,9	112	7	3165159	586	11670919	**111**
8	3.96	638,6	6.9	213,8	8	3	480	6,9	154,7	8	3	743	5	293	8	4	861	5,9	156	8	8.9	1199	8267234	100
9	82	728	84	116	9	90	545	**112**	107	9	86	638	89	**135**	9	82	433	**112**	145	9	94	1482	141	113
LO	12.55	714,93	**41**	**179**	10	11	1035	47	223	10	19	595	35	251	10	16	644	42	149	10	46	578	85	102
11	11,05	650,38	45	157	11	6,05	734,15	42	133	**11**	14	987	35	141	11	8,8	981,5	42	93	**11**	32	877	70	220
12	6.3	971,9	9,9	201,3	12	7Д	626,3	12	101	12	6.3	667,8	9	198	12	6.3	745,2	11	155	12	1232010	1245	3068416	235
13	21	939	45	285	13	14	1576	63	144	13	22	792	66	96	13	25	991	66	190	**13**	25961909	479	37848151	96
14	233	851,87	265	152	14	242	767	**231**	222	14	227	728	274	185	14	245	574	622	120	14	413611	**1161**	1310820	146
15	314	955	310	**214**	15	316	926	309	168	15	330	538	315	164	15	319	799	316	265	15	347	656	327	165
16	334.5	528,15	331	**144**	16	320	809	332	228	16	330	1014	292	197	16	314	1085	332	140	16	329	818	337	152
17	293	720	291	179	17	287	1072	281	217	**17**	297	784	289	124	17	280	586	304	149	**17**	305	677	**310**	143
18	690,4	718,31	722	186	18	658	1232	689	386	18	689	747	748	142	18	666	913	716	217	18	708	1064	804	259
19	674	852	785	146	19	717	668	722	156	19	708	705	759	272	19	721	975	739	228	19	791	699	839	254
20	685.6	1141,8	751	**197**	**20**	685,2	743,3	726	235	20	670	607	713	233	20	700	750	730	204	20	775,85	1482	839	162
21	1557	841	1570	273	21	1534	781	1573	257	21	1539	829	**1551**	213	21	1538	835	1496	362	21	1787	275	**9414435**	276
22	1600	841	1608	198	22	1575	821	1545	329	22	1503	1293	1608	262	22	1585	1196	1554	176	22	8592012	471	17513601	**121**
23	1621	992	1668	267	23	1625	1024	1677	178	23	1628	988	1615	261	23	1609	**1101**	1617	180	23	1802,9	265,8	824974338	90

Table 6.5: Results genetic algorithms (GA) generating heuristic sequencing (TS)

Source: Own elaboration

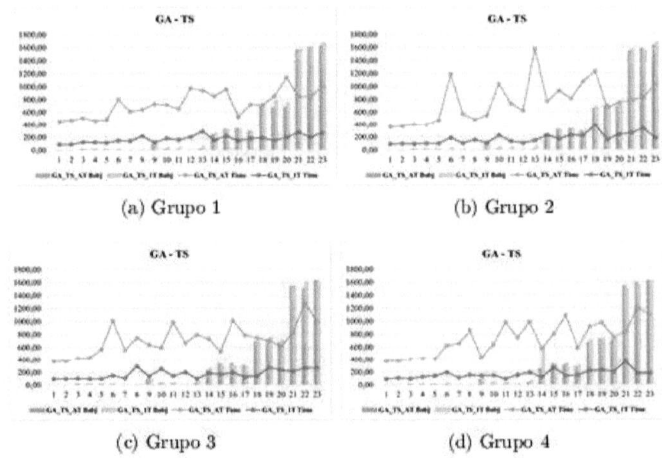

(a) Grupo 1 (b) Grupo 2

(c) Grupo 3 (d) Grupo 4

Figura 4: : Comparison of solution quality vs. computation time

Source: Own elaboration

Comparison between genetic algorithms using the nearest neighbour heuristic for sequencing generation.

In this section we will compare the algorithms that use the nearest neighbour heuristic with the difference between algorithms in the performance of the tournament. The GA_TS_AT algorithm had an average time of 734s to yield a feasible solution, while the GA_TS_1T algorithm had an average time of 166s. Furthermore, it can also be observed in table 6.5 that the algorithms in the 4 initial groups reach a similar result for the last 3 instances, i.e. the largest ones, with a difference of less than 100. As for group 5 with those same instances, a substantial difference between both algorithms can be seen, with GA_TS_1T being the least favoured.

After analysing the graphs presented in Figure 6.4 it can be noticed that these two algorithms have similar behaviours to those mentioned in the previous comparison, where the GA_TS_1T algorithm which applies the tournament only in the initial population had a lower time in all groups with respect to the algorithm where the tournament is used in all iterations. It can be seen how the GA_TS_AT algorithm reaches a high time of 1600s in group 2. In addition to this, it is observed that the quality of the solution between these two algorithms is very similar.

COMPARISON BETWEEN OPTIMISATION ALGORITHMS BASED ON THE TEACHING AND LEARNING PROCESS (TLBO)

GROUP 1	TLBO RS		TLBO TS		GROUP 2	TLBO RS		TLBO TS		GROUP 3	TLBORS		TLBO_TS		GROUP 4	TLBORS		TLBO_TS		GROUP 5	TLBO_RS		TLBO_TS	
INSTANCE	Bobj	Time	Bobj	Time	INSTANCE	Bobj	Time	Bobj	Time	INSTANCE	Bobj	Time	Bobj	Time	INSTANCE	Bobj	Time	Bobj	Time	INSTANCE	Bobj	Time	Bobj	Time
1	US	643,7	1,85	367,1	1	1,85	288,7	1,85	288,9	1	1,85	285,6	1,85	293,8	1	1,85	277,8	1,85	294,6	1	1,85	289,3	1,85	399,3
2	L85	418	1,85	389,4	2	US	**30 L**	1,85	302,9	2	1,85	287,7	1,85	309	2	1,85	285,9	1,85	296,7	2	1,85	301,4	1,85	692,5
3	15.34	425	15,34	386	3	17	297	15,34	326	3	15,34	300	15,34	310	3	16	293	15,34	319	3	**17**	292	16	397
4	**14,04**	425	14,04	387	4	14,04	312	14,04	389	4	14,04	293	14,04	329	4	14,04	504	14,04	422	4	35	308	23	311
5	16	431	14	384	5	14	**311**	14	317	5	14	484	14	315	5	14	313	17	309	5	**29**	297	20,3	398
6	12	**448**	62	753,6	6	12	465	L8	365	6	0	514	**12**	437	6	6.2	337,3	12	463	6	36	550	37	750
7	S	458	5	431	7	5	737	4	443	7	4	531	4	442	7	S	543	5	362	7	7073519	537	10577790	460
8	3	490	3	44L	8	3	377	3	372	8	3	356	4	441	8	4	360	2	487	8	9,9	349,9	11	443
9	81	705	75	**495**	9	75	872	83	470	9	81	747	74	449	9	80	983	74	811	9	**111**	521	123	546
LO	10	778	27	433	10	9.3	783,1	20	442	10	25	54 L	21	479	10	20	380	19	381	10	62	393	62	372
11	17	772	15	458	11	14	835	27	392	11	16	559	**17**	452	11	15	383	10	379	11	31	566	25	377
12	6.3	471,6	6,3	473,6	12	6.3	368,4	5,4	463,6	12	U	382,6	U	403,4	12	8,1	356,8	72	454,6	12	2115808	395	1031008	469
13	24	823	20	462	13	21	381	31	426	13	23	534	26	409	13	26	332	19	730	13	128	481	157	391
14	230	1642	218	808	14	244	**617**	204	550	14	188	757	213	669	14	198	853	209	663	14	273385	916	309010	1063
15	280	**1223**	224	**1097**	15	296	1336	243	834	15	256	830	243	**611**	15	261	1010	234	937	15	325	596	232	1097
16	261	596	248	925	16	274	857	269	953	16	310	716	268	894	16	248	765	269	953	16	292	918	291	1000
17	208	902	232	828	17	263	812	192	1193	17	230	803	197	958	17	218	1063	199	879	17	229	589	225	1260
18	609	1055	539	755	18	634	15L8	582	843	18	656	1016	569	1255	18	613	1196	575	870	18	642	1146	599	1160
19	637	2061	593	1378	19	595	771	574	1697	19	650	1736	572	1584	19	633	1116	536	1417	19	672	700	645	1920
20	660	836	636	899	20	594	1358	547	753	20	592	1354	575	885	20	630	909	553	L552	**20**	665	1393	634	1256
21	1723	1638	1363	1661	21	1675	1637	1345	1410	21	1683	897	1303	1294	21	1665	1392	1264	1311	21	1704	304	1674	301
22	1715	1795	1398	1769	22	1679	L683	1423	1558	22	**1598**	3602	L4L6	1622	22	1585	L342	L332	L542	22	**L799**	270	1685	665
23	1797	784	1460	1401	23	1612	1586	1385	2341	23	1716	**1110**	1405	1386	23	1646	1929	1399	1270	23	1734	422	1384	810

Table 6.6: Results of TLBO algorithms

Source: Own elaboration

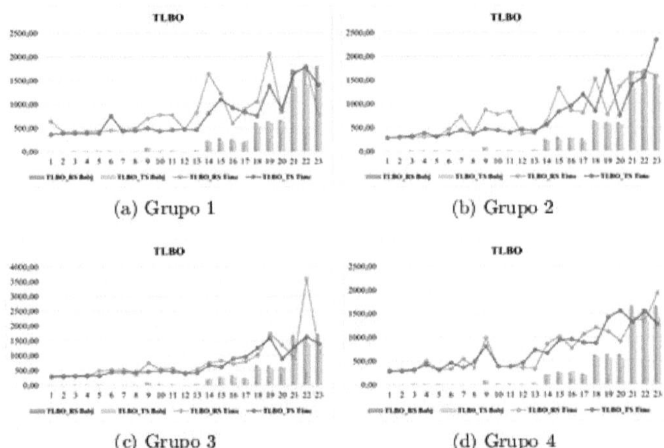

(a) Grupo 1

(b) Grupo 2

(c) Grupo 3

(d) Grupo 4

Figura 5: : Comparison of solution quality vs. computation time

Source: Own elaboration

Comparison between TLBO algorithms

In this section we compare the pair of adapted TLBO algorithms, the results obtained by the TLBO algorithm without the nearest neighbour heuristic and the other with the application of this heuristic, taking into account table 6.6 it can be said that the average time used by the TLBO_RS was 751s and the average time of the TLBO_TS algorithm was 733s, in this occasion the difference of the algorithms in terms of computational time was not so significant, however, in terms of solution quality the TLBO_TS algorithm represents better solutions.

The last mentioned is best evidenced in Figure 6.5, where the results of the TLBO_RS algorithm are always below the results obtained by the other algorithm, being more noticeable in the last 3 instances. Now, particularly in group 3, for instance 22 the TLBO_RS algorithm reaches an exaggerated time of 3602s, where a wide gap with respect to the TLBO_TS algorithm can be seen. Additionally, it can be said that only on two occasions does the TLBO_TS algorithm outperform its rival in the largest instance, attributing these occasions to groups 1 and 2.

Comparison of averages between all developed algorithms

Table 6.7 shows the average results obtained in all the groups for each algorithm proposed, both in terms of the quality of the solution (Bobj) and the computing time employed (Time).

COMPARISON OF AVERAGES BETWEEN ALL DEVELOPED ALGORITHMS

GROUP	GA RS AT		GA TS AT		GA RS IT		GA TS IT		TLBO RS		TLBO TS	
	Bobj	Tune	Bobj	Tune	Bobj	Tune	Bobj	Tune	Bobj	Tune	Bobj	Tune
GROUP 1	430,9	685,8	356.4	733,8	441,1	283,7	374,1	170,2	362,1	861,8	309,4	755,7
GROUP 2	430,4	689,1	354,2	764,6	841	239,5	367,4	173,9	350,4	804,5	304,5	744.8
GROUP 3	431,1	704,1	353,7	713,2	442,4	217,6	368,5	169,6	351,5	810,3	303,2	705,5
GROUP 4	428,5	728.4	355.4	727,7	439,6	216,7	382	165	343.9	735,8	294,3	743,6
GROUP 5	29968825	631	1711819	732	92093130	181	39742252	151	411793	545	518507	719

Table 6.7: Medians between results of proposed algorithms

Source: Own elaboration

From the table above it can be said that the changes made for group 5 significantly affect the quality of the solution of all the algorithms, due to the restricted capacity that was generated in this group, with the TLBO algorithms showing a better solution than the others. With respect to time, it is worth mentioning that the genetic algorithms that include the tournament only in the initial population outperform the others, even more so the GA_TS_1T algorithm, which on average does not take more than 174s to find a feasible solution.

It should be added that the algorithm that yields on average the best solution in all groups is the TLBO_TS algorithm, except for group 5, where the other TLBO algorithm outperforms it, so it can be said that the heuristics have a positive influence on the decision making. However, among all the algorithms there is not a very high gap in the quality of the solution for the 4 initial groups, all are approximately between 300 and 450, except for the case of the GA_RS_1T algorithm in group 2.

CHAPTER 7

CONCLUSIONS AND FUTURE WORK

7.1. Conclusions

This work focused on the development of solution methods for the integrated problem of batch sizing and sequencing of production on machines working in parallel with sequence-dependent *setups*. Six hybrid methods were developed in which the first part based on metaheuristics was in charge of determining the allocation of the batches to the machines, then the sequencing was determined and a last part, which is the exact part, was in charge of the batch sizing.

Of the proposed methods, 4 were based on the theory of biological evolution (Genetic Algorithm) where a parameterisation was carried out through a design of experiments, in order to identify the parameters that best fit the problem. The other 2 methods were based on the simulation of a classroom (TLBO Algorithm).

From the results and graphs shown in the previous section, the following conclusions can be drawn:

- The best algorithm adapted from the biological evolution (GA), in terms of computational time and response quality was the algorithm where the tournament was implemented only in the initial population and where the nearest neighbour heuristic was applied, however, it should be noted that the algorithm where the heuristic was used but with the tournament in all iterations, presented better solutions with the difference that in time the algorithm mentioned above, sometimes improved by 300 % and the response quality was only 18.41 % away on average.
- The best algorithm adapted from the simulation of a classroom environment (TLBO), in terms of computational time and response quality was the algorithm that used the adaptation of the heuristic, this algorithm took on average 755.75s to give a feasible answer, differing from the TLBO algorithm without the heuristic by 14%. This algorithm approaches and even reaches the optimum for small instances, moving away from the optimum for medium-sized instances and being slightly less efficient with large instances.
- In short, it can be said that the results produced by the 6 algorithms proposed with respect to the results shown by the exact CLSD model, reach the optimum in the instances with small scenarios, specifically for instances 1 to 5 in groups 1, 2, 3 and 4, highlighting the algorithms GA_RS_AT and TLBO_RS which reached the optimum also in instance 6, for group 5 all reach the optimum only in the first 2 instances, but show a great approach in the other small ones.

It should be noted that the main objective of this work was the comparison and/or adaptation of different approximate methods to this problem. It can be highlighted that the implementation of the nearest neighbour heuristic for the generation of the sequencing significantly helped the computation time used by each algorithm, however, the algorithms that did not use this variation presented slightly better results in the large instances.

Consequently, it is emphasised that the TLBO algorithm with the implementation of the heuristic is much closer to the solutions in the large instances, but the computation time is higher than the computation time employed by the genetic algorithm with the tournament only in the initial population and the implementation of the heuristic, this algorithm obtains better results in terms of computational time without leaving aside the quality of the solutions

in small and medium-sized instances. However, in this work a greater importance was given to the computational time with feasible solutions.
Finally, it is concluded that one of the best algorithms proposed in this work is the TLBO algorithm, knowing that with the heuristic of the nearest neighbour it saves more computational time, highlighting in the same way the genetic algorithm with the tournament only in the initial population and the heuristic sequencing, It is worth noting that in the small instances it reached the optimum, while in group 5 this algorithm in the large instances began to remain in local optima and did not improve its solution for more than an hour.

7.2. Future Work

As possible future work, it is proposed to tackle the same problem studied with a genetic algorithm in which both parts of the problem, sequencing and batch sizing, can be carried out, applying only the approximate method, comparing the results obtained by this algorithm vs. the results obtained in this work. In addition, it is proposed to study and/or adapt different additional algorithms from the literature to the problem and make comparisons of the algorithms already proposed vs. the new adapted algorithms.
In addition, it is proposed to further study other ways of adapting the TLBO algorithm, taking into account that in cases where a student is his or her own teacher, the teacher does not contribute anything to the *student phase,* and it is also proposed to consider the option that a student with better grades can learn from a relatively "bad" student.
Finally, the evaluation of larger instances is proposed, to verify again the performance of the algorithms with respect to the computation time and the quality of the solutions.

REFERENCES

Abdel-Basset, M., Abdel-Fatah, L., y Sangaiah, A. K. (2018). *Metaheuristic algorithms: A comprehensive review.* Elsevier Inc. Downloaded from https://doi.org/10.1016/B978-0 -12-813314-9.00010-4 doi: 10.1016/B978-0-12-813314-9.00010-4

Absi, N. (2008). *Models and methods for capacitated lot-sizing problems* (Unpublished doctoral dissertation). Springer.

Almada-Lobo, B., y James, R. J. (2010). Neighbourhood search meta-heuristics for capacitated lot-sizing with sequence-dependent setups. *International Journal of Production Research, 48(3),* 861-878.

Aurelio, M., Pacheco, C., Hamacher, S., y Almeida, M. R. D. (2015). Algoritmos Genéticos para Programação da Produção em Refinarias de Petróleo (January).

Azocar, H. O., y Hornig, E. S. (2014). Batch sizing and scheduling of a multi-product machine with setup and shortage. *Industrial Engineering Journal, 13(2).*

Bonrostro, J. A. P. (2015). *Operations research: the role of mathematics in decision-making: social, humanitarian and health applications: inaugural lecture of the 2015-2016 academic year.* University of Burgos, Publications and Institutional Image Service.

Boonmee, A., and Sethanan, K. (2016). A glnpso for multi-level capacitated lot-sizing and scheduling problem in the poultry industry. *European Journal of Operational Research, 250(2),* 652-665.

Bozer, Y. A., y Wang, C.-T. (2012). A graph-pair representation and mip-model-based heuristic for the unequal-area facility layout problem. *European Journal of Operational Research, 218(2),* 382-391.

Camelo Salinas, D. C. (2020). State of the art use of bubble geometry for educational purposes in Colombia.

Castillo, E., Conejo, A. J., Pedregal, P., Garcia, R., and Alguacil, N. (2002). Formulation and resolution of mathematical programming models in engineering and science. *Escuela Técnica Superior de Ingenieros Industriales, Escuela Técnica Superior de Ingenieros de Caminos, Canales y Puertos. University of Castilla La Mancha.*

Chen, S., Berretta, R., Clark, A., & Moscato, P. (2019). Lot sizing and scheduling for perishable food products: A review.

Coello, C. A. C. (1995). *Introduction to Genetic Algorithms* (No. 17).

Conover, W. J., Johnson, M. E., y Johnson, M. M. (1981). A comparative study of tests for homogeneity of variances, with applications to the outer continental shelf bidding data. *Technometrics,* 55(4), 351-361.

Contreras, IL. Sierra, E. A., Hernández, H. D., Hernández, N. B., and Moyotl, V. J. (2020). Intelligent assessment system to measure mathematical reasoning skills. *Revista Iberoamericana de Evaluación Educativa,* 75(1), 251-280.

Crainic, T. G., Davidovic, T., and Ramljak, D. (2014). Designing parallel meta-heuristic methods. *Handbook of Research on High Performance and Cloud Computing in Scientific Research and Education,* 260-280. doi: 10.4018/978-1-4666-5784-7.ch011.

Cruz, N. C., Redondo, J. L., Alvarez, J. D., Berenguel, M., and Ortigosa, P. M. (2016). *Applying a parallel teaching-learning optimization procedure for automatic heliostat focusing.*

Diego-Más, J. A., Ballester, V. C., and Santamarina Siurana, C. (2020). Optimisation of plant layout using genetic algorithms. Contribution to the control of activity geometry. *Dialnet, 1,*

415.
Dokeroglu, T. (2015). Hybrid teaching-learning-based optimization algorithms for the Quadratic Assignment Problem. *Computers and Industrial Engineering, 85*(March), 86101. Downloaded from http://dx.doi.org/10.1016/j.eie.2015.03.001 doi: 10.1016/ j.eie.2015.03.001
Drexl, A., y Kimms, A. (1997). Lot sizing and scheduling-survey and extensions. *European Journal of operational research, 99(2),* 221-235.
Fandel, G., and Stammen-Hegene, C. (2006). Simultaneous lot sizing and scheduling for multiproduct multi-level production. *International Journal of Production Economics, 10f (2),* 308-316.
Fernández, A. D., Velarde, J. G., Laguna, M., Mascato, P., Tseng, F., Glover, F., and Ghaziri, H.
(1996). Optimización heurística y redes neuronales, *ed. Paraninfo SA, Madrid, Spain.*
Ferreira, D., Clark, A. IL. Almada-Lobo, B., and Morabito, R. (2012). Single-stage formulations for synchronised two-stage lot sizing and scheduling in soft drink production. *International Journal of Production Economics, 136(2),* 255-265.
Fogel, A., Hsu, H. C., Shapiro, A. F., Nelson-Goens, G. C., & Secrist, C. (2006). Effects of normal and disturbed social play on the duration and amplitude of different types of infant smiles. *Developmental Psychology, /5*(3), 459-473. doi: 10.1037/0012-1649.42.3.459.
Fogel Pedroso, R. B. . L. E. D. Ñ. (2000). Infortunio, Dignidad Y Sabiduría De Sus Antiguos Pobladores . Asunción, Paraguay: Centro De Estudios Rurales Interdisciplinarios. , 2000.
Gómez Urrutia, E., Aggoune, IL. and Dauzère-Pérès, S. (2014, 9). Solving the integrated lotsizing and job-shop scheduling problem. *International Journal of Production Research, 52,* 5236-5254. doi: 10.1080/00207543.2014.902156.
Gogna, A., y Tayal, A. (2013). Metaheuristics: Review and application. *Journal of Experimental and Theoretical Artificial Intelligence, 25(A),* 503-526. Downloaded from http://dx.doi . org/10.1080/0952813X. 2013.782347 doi: 10.1080/0952813X.2013.782347
Goldberg, D. E. (1989). *Genetic Algorithms in Search, Optimization and Machine Learning.* AddisonWesley Publishing Company, Inc.
Guimarães, L., Figueira, G., Amorim, P., and Almada-Lobo, B. (2015). Modeling lot sizing and scheduling in practice. In *Operations research and big data* (pp. 67-77). Springer.
Hernández, J. O. (2019). Hybrid metaheuristic AG-RS approach for the buffer allocation problem that minimizes in-process inventory in open serial production lines. *,16,* 447-458.
Heuristics, (n.d.). *Meanings.com.* https://www.significados.com/heuristica/.
Holland, J. H. (1975). *Adaptation in natural and artificial systems: An introductory analysis with applications to biology, control, and artificial intelligence.* U Michigan Press.
Huaracha Ortega, M. (2015). Aplicación de juegos matemáticos para mejorar la capacidad de resolución de problemas aditivos en estudiantes de segundo grado de educación primaria de la ie ignacio merino.
Johnson, D. S., & McGeoch, L. A. (2018). The traveling salesman problem:. *Local Search in Combinatorial Optimization,* 215-310. doi: 10.2307/j.ctv346t9c.l3.
Karimi, B., Ghomi, S. F., y Wilson, J. (2003). The capacitated lot sizing problem: a review of models and algorithms. *Omega, 31* (5), 365-378.
Kis, T., and Kovács, A. (2013). Exact solution approaches for bilevel lot-sizing. *European Journal of Operational Research, 226(2),* 237-245.
Kumar, M., Mittal, M. L., Soni, G., y Joshi, D. (2018). A hybrid TLBO-TS algorithm for integrated selection and scheduling of projects. *Computers and Industrial Engineering,*

//^(October 2017), 121-130. Downloaded from https://doi.Org/10.1016/j.cie.2018 .03.029 doi: 10.1016/j.cie.2018.03.029

Liu, Q., Li, X., Liu, H., y Guo, Z. (2020). Multi-objective metaheuristics for discrete optimization problems: A review of the state-of-the-art. *Applied Soft Computing,* 106382.

López-Jiménez, D. F. (2017). The heuristic approach applied to problem solving in business: between method and strategy.

Maldonado, C. E. (2016). Metaheuristics and complex problem solving. *Revista Colombiana de Filosofía de la Ciencia, 16(33),* 169-185.

Marti, R. (2003). Metaheunstic procedures in combinatorial optimisation. *Matemátiques, University of Valencia, 1(1),* 3-62.

Martínez Lopez, C. (2018). A DISCRETE SQUIRREL SEARCH ALGORITHM APPLIED TO THE JOB SHOP PROBLEM WITH SKILLED OPERATORS. *Angewandte Chemie International Edition, 6(11), 951-952.,* 10-27.

Matai, IL. Singh, S., y Mittal, M. (2010). Facility layout problem: A state-of-the-art review. *Vilakshan: The XIMB Journal of Management, 7(2).*

Mesquita, R. (2017). *Pep: What is a production control plan and what is its importance in the logistics of your company.* Downloaded from https://rockcontent.com/es/blog/pcp/

Mor, B., Mosheiov, G., and Shapira, D. (2020). Lot scheduling on a single machine to minimize the (weighted) number of tardy orders. *Information Processing Letters, 164,* 106009.

Moujahid, A., Inza, L., and Larranaga, P. (2008). Topic 2. Genetic Algorithms. *Department of Computer Science and Artificial Intelligence University of the Basque Country-Euskal Herriko Unibertsitatea,* 1-33. Downloaded from http://www.sc.ehu.es/ccwbayes/docencia/mmcc/docs/t2geneticos.pdf

Moya Rodríguez, J., and Méndez, C. (2013, 11). Strength optimisation of plastic straight-toothed cylindrical gears using matlab software. doi: 10.13140/ RG.2.1.3739.1842.

Noble Ramos, V. M. (2017). Análise da aplicação de modelos de otimização linear na solução de problemas de dimensionamento de lotes e sequenciamento da produção de bebidas.

Novales, A. (2010). Análisis de regresión. *Complutense University of Madrid: Madrid, Spain.*

Pacheco Bonrostro, J. A. (n.d.). *Routing problems with time windows* (Unpublished doctoral thesis).

Padrón Cano, J. M. (2012). Solving the capacitated multilevel lot-sizing problem by applying particle swarm optimisation with local search.

Parsopoulos, K. E., Konstantaras, I., and Skouri, K. (2015). Metaheuristic optimization for the single-item dynamic lot sizing problem with returns and remanufacturing. *Computers & Industrial Engineering, 83,* 307-315.

Peralta-abarca, J. C., and Moreno-bernal, P. (2019). : 2007-1760 (print), 2448-9026 (digital). *,1760,* 25-32.

Pidre, J. C., Dorado, E. D., y Lorenzo, A. G. (2002). Application of Genetic Algorithms to the Problem of Classroom Allocation for Examinations in a University Centre. , 5-6.

Pinedo, M. (2014). Theory , Algorithms , and Systems (February).

Piñeros, J., Toscano, A., Ferreira, D., and Morabito, R. (2021). Datasets for lot sizing and scheduling problems in the fruit-based beverage production process. *Data in Brief, 35.* doi: 10.1016/j.dib.2021.106810.

Pousa, F. J. (2013). *A branch-and-cut algorithm for angtsp* (B.S. thesis). University of Buenos Aires.

Prado Bustamante, J. R. (1992). *La Planeacion y el Control de la Producción.* Downloaded from http ://zaloamati.azc.uam.mx/bitstream/handle/11191/4503/La_planeacion_y _el_control_BAJO_Azcapotzalco.pdf?sequence=l&isAllowed=y
Puris, A. Y., Hern, P. N., and Bayas, B. O. (2020). *Development of population metaheuristics for solving complex problems.*
Rao, R. V., Kalyankar, V. D., and Waghmare, G. (2014, 12). Parameters optimization of selected casting processes using teaching-learning-based optimization algorithm. *Applied Mathematical Modelling, 38,* 5592-5608. doi: 10.1016/j.apm.2014.04.036.
Rao, R. V., Savsani, V. J., y Vakharia, D. P. (2011, 3). Teaching-learning-based optimization: A novel method for constrained mechanical design optimization problems. *CAD Computer Aided Design, f3,* 303-315. doi: 10.1016/j.cad.2010.12.015
Crepinsek, M., Liu, S. H., and Mernik, L. (2012, 12). A note on teaching-learning-based optimization algorithm. *Information Sciences, 212,* 79-93. doi: 10.1016/j.ins.2012.05.009.
Sampieri, R. H. (2018). *Research methodology: the quantitative, qualitative and mixed routes.* McGraw Hill Mexico.
Sepúlveda, C., Díaz, L., and Arrieta, J. (2014). Prediction concurrency and algorithms in modelling.
Serna, M. A., and Marín, J. L. (2009). Genetic algorithms: an alternative solution to optimize the (Q; r) inventory model. Downloaded from http://200.12.180.26/ handle/10784/125.
Toledo, C. F. M., da Silva Arantes, M., Hossomi, M. Y. B., França, P. M., and Akartunah, K. (2015). A relax-and-fix with fix-and-optimize heuristic applied to multi-level lot-sizing problems. *Journal of Heuristics, 21(b),* 687-717.
Toro Ocampo Eliana M, G. E. M. (2005). HYBRID METHOD BETWEEN THE CHU-BEASLEY GENETIC ALGORITHM AND SIMULATED ANNEALING FOR THE SOLUTION OF THE GENERALIZED ASSIGNMENT PROBLEM.
UNED (2017). *Introduction to mathematical programming: Presentation.* Universidad Nacional de Educación a Distancia: Spain, http ://portal .uned. es/portal/page?_pageid= 93,47794967&_dad=portal&_schema=PORTAL&idAsignatura=31104021.
Valero, D. F. A., Sabater, D. J. P. G., Bas, D. A. O., and Díaz, M. A. (2001). Production programming/sequencing model for a flow shop system with different requirements according to stages. In *Iv congreso de ingeniería de organización.*
Vancells Flotats, J. (2002). Algorithms and programs. *UOC/Digitalia.*
Né\e¿, M. C., and Montoya, J. A. (2007). Metaheuristics: an alternative for the solution of combinatorial problems in operations management. *Revista Eia(8),* 99-115.
Villay Pereira, A., et al. (2013). *Analysis and development of the production planning and control system in a clothing company* (B.S. thesis). Universidad Autónoma de Occidente.

Appendices

RESULTS EXACT MODEL CLSD

GROUP 1 INSTANCE	Model CLSD Bobj	Time (s)	GAP	GROUP 2 INSTANCE	Model CLSD Bobj	Time (s)	GAP	GROUP 3 INSTANCLA	Model CLSD Bobj	Time (s)	GAP	SHOUT 4 INSTANCE	Model CLSD Bobj	Time (s)	GAP	SHOUT 5 INSTANCE	Model CLSD Bobj	Time (s)	GAP
1	1,85	0,11	0	l	1,85	0	0	1	1,85	0	0	1	1,85	0	0	1	1,85	0,02	0
2	1,85	0,02	0	2	1,85	0,03	0	2	1,85	0	0	2	1,85	0	0	2	1,85	0,02	0
3	15336	0,03	0	3	15336	0,03	0	3	15336	0	0	3	15336	0,02	0	3	15,55	0,52	0
4	14,036	0,02	0	4	14,036	0,02	0	4	14,036	0	0	4	14,036	0,02	0	4	22,852	0,52	0
5	13,534	0,03	0	5	13,534	0,02	0	5	13,534	0	0	5	13,534	0,02	0	5	20,301	0,52	0
6	0	0,01	0	6	0	0,02	0	6	0	0,02	0	6	0	0,02	0	6	18,513	0,02	0
7	1,98	0,03	0	7	1,98	0,03	0	7	1,98	0,03	0	7	1,98	0,03	0	7	415463,96	0,63	4,4E-06
8	0,99	0,02	0	8	0,99	0,03	0	8	0,99	0,03	0	8	0,99	0,03	0	8	3,96	0,59	0
9	68,848	0,03	0	9	68,848	0,03	0	9	68,848	0,02	0	9	68,848	0,03	0	9	68,848	0,47	0
10	0	0,02	0	10	0	0,02	0	10	0	0,02	0	10	0	0,02	0	10	14,97	0,02	0
11	0	0,03	0	ll	0	0,01	0	11	0	0,03	0	11	0	0,03	0	11	0	0,03	0
12	2,7	0,03	0	12	2,7	0,03	0	12	V	0,05	0	12	2,7	0,03	0	12	521003,6	0,67	5,5E-07
13	0	0,03	0	13	0	0,02	0	13	0	0,02	0	13	0	0,02	0	13	12554142,9	0,63	0
14	133,548	0,16	0	14	133,584	0,17	0	14	133,548	0,16	0	14	133,584	0,14	0	14	220975362	0,88	13E-05
15	110,928	028	0	15	110,928	027	0	15	110,928	03	0	15	110,928	028	0	15	57,503	336	3,4E-06
16	120,576	0,16	0	16	120,624	0,17	0	16	120,576	0,14	0	16	120,576	0,16	0	16	63,84	8,34	63E-05
17	98208	0,16	0	17	98,208	0,17	0	17	98208	0,16	0	17	98208	0,19	0	17	51,039	1,53	3,8E-05
18	141,84	10,92	9,6E-07	18	141,84	11,06	3.5E-07	18	141,84	9,73	0	18	141,84	12,59	0	18	7236	15,63	I,IE-08
19	161,116	12,78	4,6E-08	19	161,116	8,7	0	19	161,116	11,84	0	19	161,116	11,53	92E-05	19	8131	21,8	5,7E-05
20	151,68	20,72	92E-05	20	151,68	14,56	63E-05	20	151,68	1831	9,4E-05	20	151,68	1736	6,4E-06	20	75,84	14,41	0
21	330,51	50,13	0,0008	21	330,51	39,64	3,5E-O5	21	330,51	40,95	0,0008	21	330,51	46,64	0	21	179216	3600	0,0132
22	338,87	47,77	0	22	338,87	35,77	0	22	338,87	47,31	1JE-16	22	338,87	36,48	0	22	17827	360527	0,00288
23	331,65	40,36	0	23	331,65	46,88	0	23	331,65	36,06	0	23	331,65	52Д2	0	23	17822	3608,47	0,0053

Table 1: CLSD exact model results

Source: Own elaboration

Parameterisation Genetic Algorithm

The results obtained for the design of experiments are shown in table 2, where Bobj represents the objective function of the best solution obtained and Time the computation time associated with each of them.

INSTANCE	SOLUTIONS	CP	MP	Bobj	Time	INSTANCE	SOLUTIONS	CP	VIP	Bobj	Time
20	100	0,65	0Д5	750	251	21	100	0,65	0,15	1577	382
20	100	0.65	0Д5	718	245	21	100	0,65	0Д 5	1520	533
20	100	0,65	0,35	679	273	21	100	0,65	0,35	1561	531
20	100	0,75	0Д5	727	335	21	100	0,75	0,15	1540	378
20	100	0,75	0Д5	762	379	21	100	0,75	ОД 5	1610	378
20	100	0,75	0,35	743	210	21	100	0,75	0,35	1566	472
20	100	0,85	0Д5	767	285	21	100	0,85	0,15	1611	573
20	100	0,85	0Д 5	780	236	21	100	0,85	ОД 5	1561	**446**
20	100	0,85	0,35	722	398	21	100	0,85	0,35	1578	560
20	**200**	0,65	0Д5	716	696	21	200	0,65	0,15	1553	11.07
20	**200**	0,65	ОД 5	734	767	21	200	0,65	ОД 5	1551	1382
20	**200**	0,65	0,35	762	423	21	200	0,65	0,35	1548	1382
20	**200**	0,75	0,15	740	721	21	200	0,75	0,15	1564	739
20	**200**	0,75	ОД 5	725	774	21	200	0,75	ОД 5	1493	741
20	**200**	0,75	0,35	719	417	21	200	0,75	0,35	1531	743
20	**200**	0,85	0Д5	739	479	21	200	0,85	0,15	1582	967
20	**200**	0,85	ОД 5	726	791	21	200	0,85	ОД 5	1580	716
20	**200**	0,85	0,35	724	543	21	200	0,85	0,35	1534	665
20	500	0,65	0,15	732	1967	21	500	0,65	0,15	1553	2005
20	500	0,65	ОД 5	715	1970	21	500	0,65	ОД 5	1512	2649
20	500	0,65	0,35	726	1774	21	500	0,65	0,35	1545	1659
20	500	0,75	0Д5	732	1252	21	500	0,75	0,15	1557	1919
20	500	0,75	ОД 5	721	1971	21	500	0,75	ОД 5	1537	2650
20	500	0,75	0,35	743	1536	21	500	0,75	0,35	1554	2310
20	500	0,85	0,15	714	1602	21	500	0,85	0,15	1563	1655
20	500	0,85	ОД 5	740	1176	21	500	0,85	ОД 5	1545	2491
20	500	0,85	0,35	**711**	1787	21	500	0,85	0,35	1503	1892

Table 2: Parameterisation results

Source: Own elaboration

From the above, it was decided to perform an analysis of variance (AXOVA) for each of the response variables, this design of experiments was developed in R *software* version 4.1.2, the summary of the data can be seen in table 3:

Instance	Solutions	CP	MP	Bobj	Time
20 : 27	100 : 18	0,65 : 18	0,15 : 18	Min: 679,0	Min: 210,0
21 : 27	200 : 18	0,75 : 19	0,25 : 18	1st Qu: 728,2	1st Qu: 428.8
	500 : 18	0,85 : 20	0,35 : 18	Median: 1136.5	Median: 740.0
				Mean: 1142,5	Mean: 1021,9
				3rd Qu: 1553,0	3rd Qu: 1641,8
				Max: 1611,0	Max: 2650,0

The anova for the variable Bobj is shown in table 4:

Analysis of Variance Table

Response: Bobj

	Df	Sum Sq	Mean Sq	F Value	Pr(>F)
Solutions	2	3946	1973	3,8267	0,02902 *
CP	2	1444	722	1,4005	0,25679

MP	2	2099	1050	2,0357	0,14219
Instance	1	9095449	9095449	17640,8224	<2e-16 ***
Residuals	46	23717	516		

signif. codes 0 '***' 0,001 '**' 0,010 ,050 ,1 '' 1

Table 4: ANO VA with respect to the quality of solution

Source: Own elaboration

Considering table 4, it can be observed that the parameters CP and MP have no significance with respect to the variable Bobj, while the number of solutions can influence the result of this one, the instance factor is omitted as it is evident that they are instances of different sizes and therefore will give results far from each other. The independence assumption is verified in Figure 1, while the normality and homoscedasticity assumptions for this analysis are tested in table 5, using the Kolmogorov-Smirnov and Fligner-Killeen test respectively.

The Kolmogorov-Smirnov test was chosen as it is the most recommended test to verify the assumption of normality when the sample size exceeds 50 observations (Novales, 2010). The Fligner-Killeen test has been found to be one of the most robust homoscedasticity tests against deviations from normality (Conover, Johnson, and Johnson, 1981). In all the tests, the p-value is greater than the 5 % significance level, and Figure 2 shows the histogram of the residuals obtained, which resembles the bell shape of a normal distribution.

Response variable: Bobj

Type of test	**p-value**
Normality	0,1944
Homoscedasticity (Solutions)	0,3853
Homoscedasticity (CP)	0,9675
Homoscedasticity (MP)	0,2378

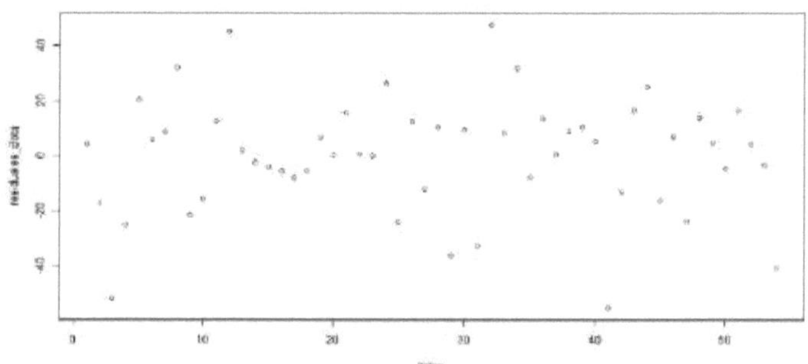

Figura 1: Graph of indenpendence of the objective function

Source: Own elaboration

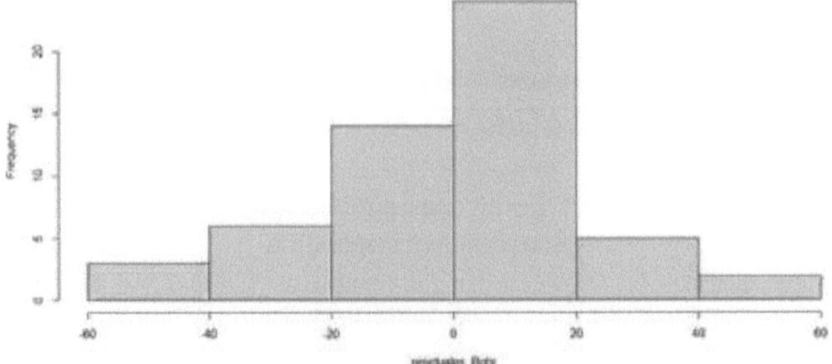

Figura 2: Histogram of the objective function

Source: Own elaboration

Table 6 shows the anova based on the variable Time.

Analysis of Variance Table

Response: Time

	Df	Sum Sq	Mean Sq	F Value	Pr(= F)
Solutions	2	22425042	11212521	198,3687	<:2,2e-16 ***
CP	2	225988	112994	1,9991	0,14706
MP	2	302951	151476	2,6799	0,07927
Instance	1	1391053	1391053	24,6101	1,002e-05 ***
Residuals	46	2600087	56524		

signif. codes 0 '***' 0,001 '**' 0,01 '*' 0,050 ,1 '' 1

In table 6 it can be seen from the p-values of each factor that the number of solutions strictly influences the computation time, contrary to the crossover and mutation probabilities. The assumption of independence is verified in Figure 3, while the assumptions of normality and homoscedasticity for this analysis can be checked in table 7, for which the Kolmogorov-Smirnov and Fligner-Killeen test were used respectively, taking into account the reasons mentioned above. In all tests the p-value is greater than the 5 % significance level, moreover in Figure 4 the histogram of the residuals obtained can be observed, which is similar to the bell shape of a normal distribution.

Response variable: Time

Type of test	p-value
Normality	0,9642
Homoscedasticity (Solutions)	0,07792
Homoscedasticity (CP)	0,9663
Homoscedasticity (MP)	0,2121

Table 7: Statistical tests of computational time

Source: Own elaboration

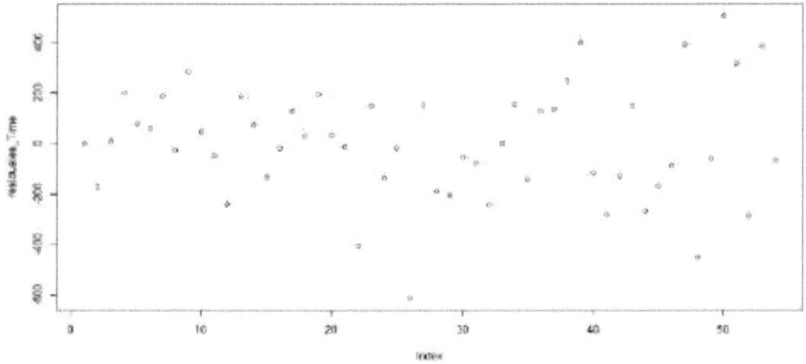

Figura 3: Computational time independence graph
Source: Own elaboration

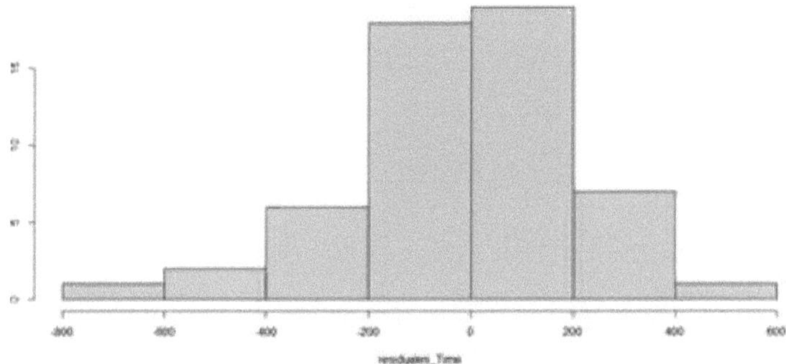

Figura 4: Histogram of the computation time
Source: Own elaboration

RESULTS EXACT MODEL CLSD

GROUP 1	Model CLSD			GROUP 2	Model CLSD			GROUP 3	Model CLSD			GROUP 4	Model CLSD			GROUP 5	Model CLSD		
INSTANCL A	Bob]	Tim s is)	GAP	INSTANCL A	Bob]	Tim e is)	GA P	INSTANCL A	Bob]	Tim e is)	GAP	INSTANCL A	Bob]	Time is)	GA P	INSTANCL A	Bob]	Time (s)	GAP
1	1.85	0,11	0	1	1.85	0	0	1	1.85	0	0	1	1.85	0	0	1	1.85	0,02	0
2	1,85	0,02	0	2	1,85	0,03	0	2	1,85	0	0	2	1,85	0	0	2	1,85	0,02	0
3	15,336	0,03	0	3	15,336	0,03	0	3	15,336	0	0	3	15,336	0,02	0	3	15,55	0,52	0
4	14,036	0,02	0	4	14,036	0,02	0	4	14,036	0	0	4	14,036	0,02	0	4	22,852	0J2	0
5	13,534	0,03	0	5	13,534	0,02	0	5	13,534	0	0	5	13,534	0,02	0	5	20,301	0,52	0
6	0	0,01	0	6	0	0,02	0	6	0	0,02	0	6	0	0,02	0	6	18,513	0,02	0
7	1,98	0,03	0	7	1,98	0,03	0	-	1,98	0,03	0	-	1,98	0,03	0	7	415463,96	0,63	4,4E-06
8	0,99	0,02	0	8	0,99	0,03	0	8	0,99	0,03	0	8	0,99	0,03	0	8	326	0J9	0
9	68,848	0,03	0	9	68,848	0,03	0	9	68,848	0,02	0	9	68,848	0,03	0	9	68,848	0,47	0
10	0	0,02	0	10	0	0,02	0	10	0	0,02	0	10	0	0,02	0	10	14,97	0,02	0
11	0	0,03	0	11	0	0,01	0	11	0	0,03	0	11	0	0,03	0	11	0	0,03	0
12	2,7	0,03	0	12	2,7	0,03	0	12	2,7	0,05	0	12	2,7	0,03	0	12	521003,6	0,67	5,5E.- O7
13	0	0,03	0	13	0	0,02	0	13	0	0,02	0	13	0	0,02	0	13	12554142, 9	0,63	0
14	133,548	0,16	0	14	133,584	0,17	0	14	133,548	0,16	0	14	133,584	0,14	0	14	220975,36 2	0,88	13E-05
15	110,928	028	0	15	110,928	027	0	15	110,928	03	0	15	110,928	028	0	15	57,503	3,36	3,4E-06
16	120,576	0,16	0	16	120,624	0,17	0	16	120,576	0,14	0	16	120,576	0,16	0	16	63,84	8,34	63E-05
17	98208	0,16	0	17	98208	0,1?	0	17	98208	0,16	0	17	98208	0,19	0	17	51,039	1,53	3,8E- O5
18	141,84	10,9 2	9,6E4J 7	18	141,84	11,0 6	ЗДЕ -07	18	141,84	9,73	0	18	141,84	12,5 9	0	18	7226	15,63	1ДДЕ4 B
19	161,11 6	12,7 8	4,6E- 08	19	161,11 6	8,7	0		161,11 6	11,8 4	0	19	161,11 6	11,5 3	92E- 05	19	8121	21,8	5,7E- O5
20	151,68	20,7 2	92.E.- 05	20	151,68	14,5 6	63E. -05	20	151,68	18,3 1	9,4E- 05	20	151,68	17,3 6	6.4E -O6	20	75,84	14,41	0
21	330,51	50,1 3	0,0008	21	330,51	39,6 4	3,5E -O5	21	330,51	40,9 5	0,000 8	21	330,51	46,64	0	21	179216	3600	0,0132
22	338,87	47,7 7	0	22	338,87	35,7 7	0	22	338,87	47,3 1	1,7E- 16	22	338,87	36,48	0	22	17827	3605 27	0,002S S
23	331,65	40,3 6	0	23	331,65	46,8 8	0	23	331,65	36,0 6	0	23	331,65	5222	0	23	17822	3608,4 7	0,0053

Printed by Books on Demand GmbH, Norderstedt / Germany